Performance and Technology

Performance and Technology

Practices of Virtual Embodiment and Interactivity

Edited by

Susan Broadhurst

and

Josephine Machon

First published 2006 by
PALGRAVE MACMILLAN
Houndmills, Basingstoke, Hampshire RG21 6XS and
175 Fifth Avenue, New York, N.Y. 10010
Companies and representatives throughout the world

PALGRAVE MACMILLAN is the global academic imprint of the Palgrave
Macmillan division of St. Martin's Press, LLC and of Palgrave Macmillan Ltd.
Macmillan® is a registered trademark in the United States, United Kingdom
and other countries. Palgrave is a registered trademark in the European
Union and other countries.

ISBN-13: 978–1–4039–9907–8 hardback
ISBN-10: 1–4039–9907–4 hardback

This book is printed on paper suitable for recycling and made from fully
managed and sustained forest sources.

A catalogue record for this book is available from the British Library.

Library of Congress Cataloging-in-Publication Data
Performance and technology:practices of virtual embodiment and
    interactivity/edited by Susan Broadhurst and Josephine Machon.
        p. cm.
Includes bibliographical references and index.
ISBN 1–4039–9907–4 (cloth)
1. Technology and the arts.   2. Digital art.   I. Broadhurst, Susan.
II. Machon, Josephine, 1979–
NX180.T4P47 2006
700—dc22   2006048581

10   9   8   7   6   5   4   3   2   1
15   14   13   12   11   10   09   08   07   06

Printed and bound in Great Britain by
Antony Rowe Ltd, Chippenham and Eastbourne

Contents

List of Illustrations	vii
Notes on Editors	ix
Notes on Contributors	x
Introduction: Body, Space, and Technology Susan Broadhurst and Josephine Machon	xv
1 Bodies Without Bodies *Susan Melrose*	1
2 Truth-Seeker's Allowance: Digitising Artaud *Steve Dixon*	18
3 Transformed Landscapes: The Choreographic Displacement of Location and Locomotion in Film *John Cook*	31
4 *Saira Virous*: Game Choreography in Multiplayer Online Performance Spaces *Johannes Birringer*	43
5 Artistic Considerations in the Use of Motion Tracking with Live Performers: A Practical Guide *Robert Wechsler*	60
6 Materials vs Content in Digitally Mediated Performance *Mark Coniglio*	78
7 Learning to Dance with Angelfish: Choreographic Encounters Between Virtuality and Reality *Carol Brown*	85
8 Kinaesthetic Traces Across Material Forms: Stretching the Screen's Stage *Gretchen Schiller*	100
9 *Sensuous Geographies* and Other Installations: Interfacing the Body and Technology *Sarah Rubidge*	112

10 Body Waves Sound Waves: Optik Live Sound and
    Performance   127
    Barry Edwards and Ben Jarlett

11 Intelligence, Interaction, Reaction, and Performance   141
    Susan Broadhurst

12 The Tissue Culture and Art Project: The Semi-Living as
    Agents of Irony   153
    Oron Catts and Ionat Zurr

13 Addenda, Phenomenology, Embodiment: Cyborgs and
    Disability Performance   169
    Petra Kuppers

14 Technology as a Bridge to Audience Participation?   181
    Christie Carson

Afterword: Is There Life after Liveness?   194
Philip Auslander

Index   199

List of Illustrations

1 Fifteen seconds, in sequence, from Rosemary Butcher's
Hidden Voices, Tate Modern, London, October 2005. The
piece lasts 15 minutes, and its sustained focus on minimal
movement in the trained dancer (Elena Giannotti) means
that viewers tend to describe the piece as 'sculptural' or
'painterly', shifting readily into the language of abstraction   5

2 In *Chameleons 4: The Doors of Serenity*, the fusion of
digitally manipulated video imagery with live
performance conjures a form of 'synergetic alchemy' to
explore and evoke Artaudian conceptions of the double,
cruelty, and the body. Clockwise from top left: Steve
Dixon, Wendy Reed, Anna Fenemore, and Barry Woods   22

3 The Chameleons Group utilise the stage and screen spaces
to conduct 'doubled' experiments with Artaud's notions
of 'physical hieroglyphs' (top 4 photos) and surrealist
conceptions of comedy and insults (bottom 4 photos) in
Chameleons 4: The Doors of Serenity   27

4 *Velazquez's Little Museum* Film Still (Courtesy of Ciné Qua
Non, Inc. Director Bernar Hébert)   40

5 *Saira Virous*, 'Blind/Inner Voice Scene', telematic dance
game, ADaPT, 2004 (Framegrab: Johannes Birringer)   54

6 *Viroid Flophouse*, showing gamers inside 'hieroglyphic
region map'. ADaPT, 2004 (Videostill: John Mitchell)   55

7 This image shows a realtime video effect linked to a
technology that responds to the touch of two dancers
(From the Palindrome opera *Blinde Liebe*, 2005; Dancers:
Aimar Perez Gali, Helena Zwiauer)   64

8 Multi-dimensional mapping   67

9 The vertical line establishes the first graphic theme in *16
[R]evolutions* (Performer: Daniel Suominen – Photo:
Richard Termine)   79

10 Traces of the performer's hands and feet leave multiple
curved white traces, a development of the white line seen
earlier in *16 [R]evolutions* (Performer: Lucia Tong – Photo:
Richard Termine)   81

11 Color and multiplicity are introduced to imply
    evolutionary change in *16 [R]evolutions* (Performers:
    Johanna Levy and Daniel Suominen – Photo:
    A.T. Schaeffer)   82
12 Catherine Bennett in *The Changing Room* (Photo by
    Mattias Ek)   88
13 Original Sketches by Colin Lombard of images of Marey's
    experimental shoes and recording instrument ca. 1872 (a)
    and of Loïe Fuller dancing 'The Lily', ca. 1900 (b). These
    sketches were inspired from images of Marta Braun (1992:
    27) and a photo by Isaiah West Taber (From Musée de
    l'Ecole de Nancy, 2002).   102
14 Optik perform *Xstasis* (2003), Montreal, Canada (Photo:
    Alain Décarie)   139
15 Optik perform *Xstasis* (2003), Montreal, Canada (Photo:
    Alain Décarie)   139
16 Elodie and Jeremiah from *Blue Bloodshot Flowers* (2001)
    (Image by Sally Trussler and Richard Bowden)   142
17 Katsura Isobe, Dave Smith and Tom Wilton in *Dead East,
    Dead West* (2003a) at the ICA, London (Image by Terence
    Tiernan)   145
18 *Disembodied Cuisine* Installation, from L'Art Biotech
    Exhibition, France 2003, The Tissue Culture & Art Project
    2003 (Photography: Axel Heise)   162
19 Victimless Leather – *A Prototype of Stitch-less Jacket grown in
    a Techno-scientific 'Body'*, The Tissue Culture & Art Project
    2004. Medium: Biodegradable polymer connective and
    bone cells (Dimension of original: variable)   164
20 Installation Still, *Body Spaces*, Petra Kuppers, © 2000   177

Notes on Editors

Susan Broadhurst is a writer and practitioner in the creative arts. She is Reader in Drama and Technology, Head of Drama Studies and Head of Performance of the BitLab Research Centre at Brunel University, West London. She is also the author of *Liminal Acts: A Critical Overview of Contemporary Performance and Theory* (London/New York: Cassell/Continuum, 1999) and *Digital Practices: A Critical Overview of Performance and Technology* (forthcoming, 2007). Her various articles include 'Interaction, Reaction and Performance: The Jeremiah Project,' which was commissioned by *The Drama Review* (MIT Press 48 (4) (December 2004): 47–57), and she is Co-editor of the *Body, Space & Technology* online journal <http://www.brunel.ac.uk/bst/>. Susan is currently working on a series of collaborative practice based research projects entitled, 'Intelligence, Interaction, Reaction and Performance', which involve introducing various interactive digital technologies into live performance, including artificial intelligence, 3D film, modelling and animation, and motion tracking.

Josephine Machon is Programme Director for Physical Theatre, and a lecturer in Drama and Performance Studies at St. Mary's College, University of Surrey, West London. On her topic (syn)aesthetics and performance, she has published various writings. Among her multimedia works, Josephine has authored 'To Deliver Us from (Syn)aesthetics', an interactive document in *Flesh & Text*, a CD ROM marking 10 years of Bodies in Flight (October 2001). She is in the process of writing *(Syn)aesthetics – Towards a Definition of Visceral Performance* (forthcoming 2007) and is co-authoring *British Physical Theatre – Practice & Practitioners* (forthcoming 2006). As a practitioner she has a broad range of experience including her current collaborations, which explore the fusion of embodied practice and the written text within *play*ful performance encounters. Josephine is Sub-Editor for the online journal *Body, Space & Technology*.

Notes on Contributors

Philip Auslander teaches Performance Studies in the School of Literature, Communication, and Culture of the Georgia Institute of Technology (Atlanta, Georgia, USA). He contributes regularly to journals such as *TDR: The Journal of Performance Studies* and is on the editorial board of that journal, *Performance Research*, the *International Journal of Performance Arts and Digital Technology*, and others. He is the author of five books, most recently *Performing Glam Rock: Gender and Theatricality in Popular Music* (2006). In addition to his work on performance and music, Auslander writes on the visual arts for *ArtForum* and other publications.

Johannes Birringer is an independent choreographer and media artist. As artistic director of AlienNation Co., an ensemble based in Houston (www.aliennationcompany.com), he has created numerous dance-theatre works, video installations, and digital projects in collaboration with artists in Europe, the Americas, and China. He is the author of several books, including *Media and Performance: Along the Border* (1998), *Performance on the Edge: Transformations of Culture* (2000), and *Dance Technologies: Digital Performance in the 21st century* (forthcoming). Founder of the Interaktionslabor Göttelborn in Germany (http://interaktionslabor.de), he is currently Professor of Drama and Performance Technologies at Brunel University, West London.

Carol Brown is a choreographer, performer, and writer. Originally from New Zealand, she has been based in London since 1995. Her company, Carol Brown Dances, is renowned for its cross-art form works for theatre, installation, and screen, and has performed throughout the world. Carol has one of the first doctorates by practice from the University of Surrey in England and is a senior research fellow at Roehampton University where she is part of the Centre for Dance Research. Her current cycle of work is concerned with new bodily perspectives and spectral effects emerging at the confluence of technology and mythology. Recent awards include a Ludwig Forum International Prize for Innovation (2003) and a NESTA Dream Time (2004).

Christie Carson is a senior lecturer in the Department of English at Royal Holloway University of London. Prior to this appointment, she worked as Research Fellow in the Department of Drama and Theatre at Royal Holloway and was Director of the Centre of Multimedia Performance History there from 1996 to 2003. She is the co-editor of *The Cambridge King Lear CD-ROM: Text and Performance Archive* (Cambridge, 2000) and author of 'King Lear in North America', an article on this CD. Dr Carson has also completed a large AHRB-funded research project entitled *Designing Shakespeare: An Audio-visual Archive, 1960–2000*, which documents the performance history of Shakespeare in Stratford and London. This database is freely available on the web at www.ahds.ac.uk/performingarts/designing-shakespeare.

Oron Catts is an artist/researcher and curator, and co-founder and Artistic Director of SymbioticA, The Art & Science Collaborative Research Laboratory at The School of Anatomy and Human Biology, University of Western Australia. Catts founded the Tissue Culture and Art Project (TC&A) in 1996. The Tissue Culture and Art Project is an ongoing artistic research and development project into the use of tissue technologies as a medium for artistic expression.

Mark Coniglio is a composer/media artist and creates large-scale performance works that integrate music, dance, theatre, and interactive media. With choreographer Dawn Stoppiello he is Co-founder of the New York City-based dance theatre company Troika Ranch, which for the last decade has been created numerous multidisciplinary performances. Coniglio is particularly interested in creating custom interactive systems that allow the movement or vocalizations of the performers to generate and/or manipulate video, sound and other digital media. Coniglio is also the creator of real-time media manipulation software *Isadora®*, which is used by artists worldwide to realize their interactive performances and installations.

John Cook, PhD, is an independent scholar located in Istanbul, Turkey. He currently runs his own performance and consulting company, Perishable Produce, dedicated to preserving the environment through sustainable practices and utilizing alternative energy resources. Dr Cook has organized two international dance film festivals and taught courses in the Fine Arts at Bosphorus University and Koc University in Istanbul, Ohio University and DePaul University, and Columbia College and Northwestern University in Chicago.

Steve Dixon is Professor of Performance and Technology at Brunel University, where he is also Head of School of Arts, and Deputy Director of the BitLab Research Centre. His creative practice includes international multimedia theatre tours as Director of *The Chameleons Group* (since 1994), two award winning CD-ROMs, interactive Internet performances, and telematic arts events. He has published extensively on subjects including performance studies, film theory, digital arts, robotics, artificial intelligence, and pedagogy. He is Associate Editor of *The International Journal of Performance Arts and Digital Media*, and Co-director of *The Digital Performance Archive*.

Barry Edwards is Director of Optik. Directing credits also include new writing, classic adaptation, object and puppet theatre, experimental film, and musicals. University experience includes Concordia Montreal, Sofia, London University of the Arts, Central School, and Brunel University. His work has been produced in Alexandria, Berlin, Sao Paulo, Montreal, Helsinki, and many other cities worldwide.

Ben Jarlett, MSc BEng, is a research assistant at Brunel University and undertaking PhD research at Bath Spa University into programming and interface design for computer-based improvisation and live performance using Max, MSP, and Jitter. He has been performing with Optik since 2000.

Petra Kuppers is Associate Professor in the English Department at the University of Michigan, Ann Arbor, where she teaches in performance, cultural, and disability studies. She is the author of *Disability and Contemporary Performance: Bodies on Edge* (2003), *The Scar of Visibility: Medical Performances and Contemporary Arts* (in press) and *Community Performance: An Introduction* (forthcoming 2007). This last volume has a companion edition, *Community Performance: A Reader*, which she is co-editing with Gwen Robertson (2007). Petra Kuppers is also the Artistic Director of The Olimpias Performance Research Projects (www.olimpias.net).

Susan Melrose is Professor of Performance Arts and Research Convenor for Performing Arts, in the School of Arts, Middlesex University. After completing her doctoral research in performance analysis at the Sorbonne Nouvelle in the early 1980s, she established and ran post-graduate profession/vocation-linked theatre and performance studies courses at Central School of Speech and Drama and Rose Bruford

College, London, before taking up her professorship at Middlesex. She is widely published in the fields of performance analysis, performance writing, and critical semiotics. Her Semiotics of the Dramatic Text, for Macmillan, appeared in 1994. In 2005 she co-edited *Rosemary Butcher: Choreography, Collisions and Collaborations* for Middlesex University Press. A number of presentations and keynote papers are web-published at http://www.sfmelrose.u-net.com.

Sarah Rubidge is a practitioner-scholar whose work has been published in journals and books since 1981. She is a choreographer and digital installation artist and currently Reader in Digital Performance at the University of Chichester. The focus of her artistic work lies in the dialogue between the body, movement and new technologies, in particular in interactive installations. Sarah is especially interested in developing installation spaces which are read through the haptic/kinaesthetic senses. She is also interested in developing performative installation spaces in which participants' movements become an integral element of the installation event. Sarah is on the editorial board of Dance Theatre Journal and Body Space Technology.

Gretchen Schiller explores physical and visual vocabularies by conjugating qualitative movement with video, performance, and interactive participatory environments. Her research includes aesthetic mapping, multiple video projection and non-linear dynamic forms. Her work has toured in Canada, the United States, England, Germany, and France. She has recently joined the School of Arts at Brunel University after teaching at the University of Montpellier for seven years. She received her BA from the University of Calgary, Canada, MA from the UCLA University of Los Angeles, California, and PhD from the University of Plymouth, UK. She is the artistic director of Mo-vi-da, http://www.mo-vi-da.org, which supports the creative research in the areas of media dance.

Robert Wechsler is a performing artist and researcher in the use bio-sensor and video-based motion tracking. He is artistic director of Palindrome, an award winning performance group specialized in interactive performance. A Fulbright Fellow, he is the author of articles in *Leonardo, Ballet International, Dance Magazine,* and *Dance Research Journal.* He was Designer and Head of the Masters Degree Programme in Digital Performance at Doncaster College from 2004 to 2005. In 2006, he worked with composer Dan Hosken and choreographer Helena Zwiauer on *A Human Conversation,* an ongoing project to explore the relationship between human movement and speech. Mr Wechsler leads seminars and workshops in motion tracking at institutions worldwide.

Ionat Zurr is an artist in residence in SymbioticA and a co-founder of the Tissue Culture and Art Project. She has studied photography and media studies, and is currently a PhD researcher looking at the ethical and philosophical implications of biological art.

Introduction: Body, Space, and Technology

Susan Broadhurst and Josephine Machon

> Habit expresses our power of dilating our being in the world, or
> changing our existence by appropriating fresh instruments . . .
> the body is our general medium for having a world.
>
> – Merleau-Ponty (1962: 143–6)

In *Performance and Technology: Practices of Virtual Embodiment and Inter-activity*, we offer a collection of writings from international contributors who specialise in digital performance practices. These performances cross and blur the boundaries between dance, film, theatre, installation, sound, and biotechnology. They also employ a diverse range of new technologies.

Motion tracking is one such technology that currently uses magnetic or optical motion capture and has been utilised widely in performance and art practices. It involves the application of sensors or markers to the performer or artist's body. The movement of the body is captured and the resulting skeleton has animation applied to it. This data-projected image then becomes some part of a performance or art practice (see Chapters 5 and 6). Another technology highlighted in these practices is artificial intelligence, where the challenge is to demarcate the delimited human body from an artificially intelligent life form, such as Jeremiah, the avatar (see Chapter 11). In other words, there is an emphasis on the play between the human and technological exchange in such interaction.

Again, there is a proliferation of performances that utilise electronic sound technology for real-time interaction. A performance group who explores the use of this technology is Optik, who have performed at various national and international venues and now prioritise the use of digitally manipulated sound in their movement-based performance (see Chapter 10). Furthermore, in recent years, there has been a preponderance of art works that incorporate biotechnology within their creative experimentation that carries with it challenging ethical implications. Such ventures are commonly referred to as 'Bioart'. The Tissue, Culture and Arts Project are such a group, whose tissue engineering exploration is integral to their art installations, resulting in works of varying geometrical complexity, thereby creating a living 'artistic palette' (see Chapter 12).

In our opinion, such works present innovation in performance and theory, being at the cutting edge of creative and technological experimentation. As such, we identify certain features that are quintessential to these new practices. One such prominent feature is the absolute centrality of the digital, even though in the various artworks and performances presented here, there is a diversity of technologies employed. Another exemplary feature that runs throughout this collection is an emphasis on the corporeal in terms of both performance and perception.

Consequently, in our opinion, such quintessential features demand a new mode of analysis and interpretation which foregrounds and celebrates the inherent tensions between the physical and the virtual. This simultaneously both morphs and extends the performing body, thus engendering an altered corporeal experience.

The readings proffered in this collection stress the emotive, the intuitive, the ludic, and the sensate, since in many art forms the body is primary and yet transient. Thus, the immediacy of the physical/virtual body, *including* its corporeal reading, is made the focus of interpretation. This acknowledges the complexity of the immediate relationship between audience and performance, thereby celebrating the 'total' appreciation of these digital *Gesamtkunstwerke* (collective works of art) by the audience. In short, due to this preponderance of technological performances, we suggest a new aesthetics is created that demands new approaches to appreciating such works.

By surveying and interrogating various aspects of performance and technology from our contributors, we have endeavoured to create a forum of debate that addresses these concerns. In editing this collection, we have attempted to create a flow of interactivity and exchange between the papers themselves. Whilst the reader may detect an apparent trajectory in the ordering of the chapters, we would like to stress that we see an overarching interaction between *all* the various perspectives and approaches offered as much as each being viewed as a discrete entity in itself.

Susan Melrose opens with a discussion on Slavoj Zizek's re-reading of Gilles Deleuze, with added references to Brian Massumi's notion of 'affect' and Katherine Hayles' notion of the 'resistant materialities of embodiment'. It also brings into question the concept of 'hypotyposis', that is the indirect presentation of a concept within a piece of art that connects the intuition with the understanding. Melrose explores how the notion of virtual time remains relatively undertheorised though simultaneously it remains central to technology and performance.

For Steve Dixon, the notion of artistic 'truth' has been undermined by recent theory, yet a search for some form of truth still underlies many performance artists' work. The Chameleons Group is used as a case study: a company fusing digitally manipulated video projections with live theatre to follow Artaudian and Surrealist ideas, and to update their very particularised notions of *truth* for a contemporary audience. The melding of the live and the virtual, Dixon argues, is to provoke a type of alchemical reaction for both actors and audiences that can awaken and ignite transformative Artaudian concepts of 'the double', the 'actor in delirium', and 'physical hieroglyphs'.

According to John Cook, although choreographic and cinematic collaboration began over 110 years ago, little has been written on the transformative kinetic interplay between the two. Cook examines works by various North American and European choreographers and filmmakers exploring the ways new technologies make it possible for the choreographer to displace, without replacing, the presence of the performer – moving the performer and viewer to a new place/si(gh)te at the same time making the 'place' move. The choreographic language speaks to, and with, the physical location; a location that now seems more real or imbued with more 'depth', than the theatrical setting of a theatre. In doing so, the somatic experience of the performers and of the spectators expands into a new realm of techno presence.

The alternative multimedia event *Saira Virous* (2004) is discussed by Johannes Birringer. He explores the idea of choreography as open process and emergence, understood in terms of interactive real-time game performance in a multi-site telepresence constellation. It is a collaborative creation process, which involves scripting in a dramaturgical/notational sense and in the sense of programming a responsive environment (the 'game engine'). The project's conceptual development owes much to recent cross-currents in game design and interactive media art practices, while its focus on physical performance inside programmable environments extends the author's choreographic work in the field of dance and technology.

Robert Wechsler is a long-time user of the EyeCon system, developed by Frieder Weiß, with whom he co-directs Palindrome Intermedia Performance Group. While reviewing the features and applications of EyeCon, he draws many conclusions that could be applied to motion tracking and interactive performance in general. However, as Wechsler states, it is important to 'think of interaction primarily as a psychological phenomenon, rather than a technical one'. Cameras and computers are not needed since 'interactivity is simply the instinctive back and forth of energy' which occurs during communication and creative practices. Mark Coniglio proposes that there are two basic approaches to making media intensive performance: materials-driven and content-driven. The former, he suggests, is primarily focused on digital materials and the performative contexts they indicate, the latter uses digital materials in an attempt to communicate specific narrative ideas. By considering the arc of other métiers that have adopted new technology in the past, he posits questions about the field's current state-of-the-art; how these materials can and/or should be used, and the future implications that there may be for those who embrace either a materials- or a content-driven approach.

Carol Brown focuses on the refolded space of choreography in the digital age where she suggests a haunting virtuality is discovered. Brown proposes specific gestures through which the staging and dramaturgy of dance performance can be re-envisioned through the incorporation of virtual dimensions. She questions the new modalities of performance that emerge from the mutual imbrication of flesh and data in environments which blend real and virtual dimensions. The tension between possessing and inhabiting virtual forms in embodied interface design is given a critical framework for understanding within the legacies of performance practice.

According to Gretchen Schiller, at the end of nineteenth century, mechanical and electronic techniques were developed to translate and transfigure bodily movement into new visual forms and kinaesthetic metaphors. Schiller examines how such techniques introduced by performer Loïe Fuller (1862–1928) and scientist Etienne Jules Marey (1830–1904) re-appear in a contemporary movement-based interactive installation entitled *trajets* (2000). She draws a parallel between Fuller's and Marey's seminal contributions and the choreomediated techniques employed in this performance to ellicit the kinaesthetic experience of the visiting public.

Sarah Rubidge discusses the range of theories which have impacted on the development of her installations, including those of Henri Bergson, Gilles Deleuze, and Felix Guattari. She interrogates the notion that an interface between the more subtle aspects of the human body (such as the autonomous physiological systems) with complex responsive systems devised through and with new technology can be illuminated for both the artist and the participants in the installation events.

The journey taken by Director Barry Edwards and Sound Artist Ben Jarlett to build a collaborative and responsive digital sound-scape for the performance work of the company Optik is discussed in Chapter 9. This begins with the early experiments in Internet-linked sound and video transmission, and goes on to chart the subsequent experiments with live capture and granular synthesis. In interrogating their experiences, their discussion includes technical details relating to the audio processing techniques used and collaborative strategies developed that integrate live musicians, teams of digital sound artists, and live performers.

Susan Broadhurst proposes that new liminal spaces exist where there is a potential for a reconfiguration of creativity and experimentation. These spaces are liminal in as much as they are located on the 'threshold' of the physical and the virtual, and as a result tensions exist. It is suggested that it is within these tension-filled spaces that opportunities arise for new experimental forms and practices. Her practice-based project titled 'Intelligence, Interaction, Reaction and Performance' is a series of performances and installations that analyse and explore the interface between physicality, digital interactivity, and AI technology in contemporary art and performance practices. This research investigates the aesthetic potential of digitised technology for arts practice as exemplified by her performance work.

The focus for Oron Catts and Ionat Zurr from Tissue Culture and Art Project (TC&A) is the artistic practice and resultant ethical and moral dilemmas of tissue engineering. In the last few years, they have been applying tissue-engineering principles for the purpose of artistic expression. These entities (sculptures) blur the boundaries between what is born/manufactured, animate/inanimate, and further challenge our perceptions and our relations towards our bodies and constructed environment.

In a consideration of 'cyborgs and disability' in performance, Petra Kuppers focuses on the morphing body of Aimee Mullins as a presence in art photography, performance film, and advertising. As a real-life cyborg and disability celebrity, Kuppers shows how Mullins' presentation allows for a complex and multi-levelled engagement with discourses of beauty, difference, and representation. Kuppers goes on to reflect upon her own collaborative performance practice, *Body Spaces* (2000), which was fuelled by an interest in the relationship between bodies, addenda, and space. This shared architectural/environmental framework is normalised for bodies that fit its criteria. However, 'different bodies' become highly 'visible' and 'tangible'.

Christie Carson shows how technology can be a 'bridge to greater audience participation'. She examines the way in which the relationship between the performer and the audience has been altered, not only in those productions which use technology directly as part of the performance, but in all theatre due to new expectations of interaction set up by digital communication. Looking at developments in the theatrical establishment in response to, on the one hand, experimentation in the smaller theatres and on the other hand, to changing expectations in the audience, she demonstrates how far-reaching the changes which our technological society has instigated have been. Carson shows how digital communication technologies connect audiences to a theatrical community which extends beyond the theatre doors. In doing so, she suggests that technology can break down the social and creative barriers between the audience and the onstage event, encouraging the democratisation of the theatrical experience.

Finally, within his 'Afterword', Philip Auslander develops his own argument regarding the nature and experience of 'liveness' and 'after liveness' in performance. As a result, he generates debate between practitioners and academics regarding these new movements in (and between) bodies, spaces, and technologies.

Reference

Merleau-Ponty, Maurice. (1962). *Phenomenology of Perception*, trans. Colin Smith, London: Routledge.

1
Bodies Without Bodies

Susan Melrose

> The simplest non-biological instance of spontaneous correlation between the probabilities of events is the behaviour of materials *near phase transitions*.
>
> – M. DeLanda (2002)
>
> Philosophy is drawn to the question of difference, that is, to the immersion of difference in and the production of difference by duration. Duration is difference, the inevitable force of differentiation and elaboration, which is also another name for becoming.
>
> – E. Grosz (2000)

Starting points

A slender volume was published by Gilles Deleuze in French, in 1970, with the curious title: *Spinoza: Philosophie pratique*. The book was revised and expanded, and republished under the same title, again in French, in 1981. Curiously – as far as I am concerned – it was not translated into English (R. Hurley), and published by City Lights Books, until 1988. Do history (of writing) and time (or timing) of translation and republication *matter*? In English-language translation, Deleuze's *Spinoza: Practical Philosophy* . . . appeared after *Anti-Oedipus* (1984) and *A Thousand Plateaus* (1987), both of which were co-written by Deleuze and Guattari. On this basis, Deleuze's text 'on' the Seventeenth-century philosopher, Spinoza, might seem to have reached an English-language readership less on the basis of its own merits, and rather more because of the then notorious collaboration with Guattari. Many readers today will be aware that it was in his *Spinoza: Practical Philosophy*, at the end of

1

the 1980s, that Deleuze seemed to address a particular observation to English-language readers: 'Spinoza offers philosophers a new model: the body. He proposes to establish the body as a model: "We do not know what the body can do"'. 'This declaration of ignorance', Deleuze added, 'is a provocation' (p. 17).

Now, writing today as something other than a professional philosopher, I want to expose my own ignorance: I do not know what *performance* writers mean more generally, when they use a number of terms, key amongst which is that same 'the body'. I am more or less persuaded, by the same token, that Spinoza's 'the body' is largely incommensurable, despite the urgings of common sense, with 'the body' referenced by recent dance writers concerned with 'new work'. Perhaps at this precise point I should also point out that whereas professional philosophers tend to remain within the relatively comfortable universe of writing, performance practitioners and performance writers tend, by definition, to operate productively, in significant part, outside of writing, in areas for which no easy fit with the *orders* of writing is necessarily available.

According to Zizek, writing around 2003/2004 – with the considerable advantage of hindsight – Deleuze's collaboration with Guattari afforded the former an 'easy escape from . . . his previous [pre-Guattarian] position'. An escape was needed, according to Zizek, because of the deadlock, in Deleuze's single-authored writing, between two key notions: first, a 'logic of sense', and second, a 'logic of becoming' (Zizek, 2004: 21). The Deleuzian 'logic of sense', according to M. DeLanda's 'virtual philosophy' (2002), was a matter of an '*immaterial* becoming' (pp. 107–8; my emphasis); of multiplicities which are actually 'causally sterile entities' – perhaps because those multiplicities remain of the order of the system, within which they turn, and turn again. The Deleuzian 'logic of becoming', according to the same writer, is, on the other hand, a matter of the *production* 'of beings', enabled when 'metric or extensive properties' emerge, in a single process, 'in which a virtual spacetime progressively differentiates itself into actual discontinuous spatio-temporal structures' (p. 122).

My own interest lies in attempts to grasp the production of 'new work' discursively as distinct from the interpretation of the already-made, from the perspective of spectating. I am going to understand 'production', here, to be purposeful, objective-driven, focused on a future ('output'). Hence the emergent, here, is time-marked as well as time-dependent: it is a matter of operations within the 'performance economy', where that economy itself presses in on processes as though from their outside. I am supposing that the production of beings necessitates a producer or producers, as well as – in the fields which interest me here – disciplinary (even more than interdisciplinary) processes of production. I propose in addition to characterise that production *politically*, at this point, as expert or professional, and to add to those qualifiers the term *existential* (Osborne, 2001) in order to signal that such production ('making new work') is constitutive to that producer's personal-professional being. To the extent that the 'new work' concerned aspires to, and tends to, involve *expert invention*, even innovation (I am thinking here of the work of the Wooster Group, or Robert Wilson, or Ariane Mnouchkine, but also of Complicite, of desperate optimists, of DV8, and of Wayne McGregor, more locally), I propose in addition to qualify such production processes in terms of their orientation to 'qualitative transformation' (Massumi, 2002): that is, to the production of what its makers identify as 'better' work.

'New' work's time

If I transpose elements of Manuel DeLanda's 'virtual philosophy' (2002) to a significantly different context of production (making 'new work'), which also entails 'stepping out of' philosophical writing, in order to contemplate its other, then a 'Deleuzian' logic of (performance-productive) becoming would seem to involve, *for an expert practitioner*, processes allowing the 'progressive differentiat[ion]' from 'a continuous virtual spacetime' (e.g. 'devising'), on the basis of which 'actual discontinuous spatio-temporal structures' are arrived at, however temporarily.

In the terms I have begun to set out, such differentiation – in 'making new work' – is driven by certain expectations, including a timely outcome ('the show'), which is time-determined or determining a number of times over: typically the actual production deadline tends to fold back over, and inform, detailed decision-making; second, the specificity of the performance 'event' determines the play of time within the devised work; but third, time conditions the exposure and display of performance elements themselves, in performance-genre-specific terms. Finally, where the signature practice of a choreographer is concerned, it is worth observing that 'the show itself' tends to function as a pause in that practitioner's ongoing processes: it represents a momentary instantiation (Knorr, 2001), and not that performance maker's 'thing itself'. Each of these involves 'measure', both quantitative and/or qualitative.

You may well have noted that my references to performance thus far tend to be theatrical, in terms which are arguably 'pre-posthuman'. It might well be observed that I am recuperating notions from DeLanda's 'virtual philosophy', on a largely metaphorical basis and/or via the operations of my writerly expert intuition, into traditional frames of reference. I have wondered whether I have practised this recuperation because I am actually 'cognitively-mapped' (Jameson, 1991) in terms of older technologies and their apparatuses. If that were the case, then it might also be true of a generation of performance-writers, who similarly tend to measure and recuperate aspects of the discourses/practices of technological change, in terms characterised by older logics. On the other hand, to return to my starting point, this issue of cognitive mapping in terms of older technologies would also then apply to the case of Spinoza, writing in the 1670s, and equally to Deleuze, writing in the late 1960s. On this basis, I am obliged once again to signal my ignorance: I do not know 'what "the body" [might have meant]', in terms of older 'technologies'. In common-sensical terms, 'a body' is 'a body'; yet I am not prepared to suppose that common sense provides the best bases for contemplating what is going on in the making of 'new work'.

Despite my acknowledgement that I might be recuperating the 'virtual philosophy' of DeLanda, via an expert-writerly *sense* of 'fit' at the level of (conceptual) 'equipment' (Rabinow, 2003) or apparatus, I do nonetheless want to pursue the notion of 'metric properties' in terms of judgement and 'measure'. Measure can apply both in terms of the technical (metrisation or quantisation) and in terms of judgements of taste and value (Bourdieu, 1977), which condition the emergence and evaluation of 'bits' of 'new work'. It is by the measure in both senses, of 'actual discontinuous spatio-temporal structures' (such as 'character' or 'performer', or 'ending'), that 'we' identify 'them' as such. Measure, in this specific sense, where the production of 'new performance' material is concerned, seems to be linked to a capacity, in emergent material, to sustain a certain *duration* of (inquisitive) regard. Far from swiftly achieving 'fit' (itself a matter of measure), that emergent new performance material might well require, of the 'plane of immanence' (Deleuze and Guattari, 1994) specific to the discipline (e.g. 'new dance', and what makes it 'dance'), that it maintain its disciplinary identity, while expanding to accommodate the new. On this sort of basis, the 'plane of immanence' can be identified as dynamic but also speculative. Its 'measure', equally, tends to be a matter of relation (i.e. it is relationally determined) in expectation of, as well as in the event of, spectating.

In other words, measure involves a complex relation between participants to decision-making, within the economy of expert performance production. 'New work', in the performance disciplines which interest me here, is necessarily performative, in the sense that its impact, where

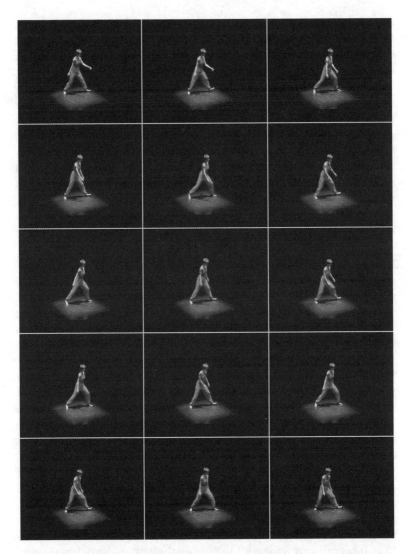

Figure 1 Fifteen seconds, in sequence, from Rosemary Butcher's *Hidden Voices*, Tate Modern, London, October 2005. The piece lasts 15 minutes, and its sustained focus on minimal movement in the trained dancer (Elena Giannotti) means that viewers tend to describe the piece as 'sculptural' or 'painterly', shifting readily into the language of abstraction

disciplinary practices are concerned, depends upon its capacity to target, to draw, and to hold the other's gaze. Now, 'targeting', 'drawing', and 'holding', metaphorical though these terms are here, tend to be a matter of time and timing. There is no visual 'focus', in the performance *event*, which is not time-governed. Its measure is bound up in its capacity to capture and to maintain attention over time where a '*per*-formance event' is concerned. The logic of performance-production is a matter of multiple and interlocking instances of measure, conjured (in terms of Charles Peirce's late-nineteenth-century semiotics) by someone, for someone, somewhere, and to some end or ends; that measure, as I proceed to demonstrate, is likely to be dependent in part at least, at all stages, on expert-intuitive processing.

Timely reconfigurations

It seems to be 'targeting', 'drawing' (attention), and 'holding', over time, that allows 'new work' to seem, to a spectator, *to do more* than it actually does (Bourdieu, 1977). Performance tends to evoke, rather more than to 'represent'; on this basis, there is something curious about whatever is identified, in the show, in terms of 'the body'. 'The body', *in the event*, is theatrical: it is understood to be multi-dimensional, and it consistently signals, in performance-disciplinary terms, 'more than' can be seen.[1] Engaging with this 'more than' is, once again, a matter of judgement, of measure; and judgement tends to operate contrastively, in time. 'New work' is theatrical, in terms of its capacity to make a spectator look twice and look carefully – both of which are time-based activities; and on the basis of that twice-looking, what *follows* (again, time is of the essence) is a spectator's further engagement with the (measure of) the experience.

In the disciplinary, performance-expert terms I have begun to set out above, measure, both quantitative and qualitative, remains a marker of disciplinary-identity, of 'immanence', of expertise, of 'event', and – indeed – of that most curious unanalysed element of expert performance-making, which is casting (or its equivalent). Measure *takes time*: it begins early and tends to operate through multiple instances of expert intuition with the future work in mind. It is determining, many times over, even where a practitioner group has defined itself as 'challenging', or its practices as 'cutting edge'. Let's say so, not least because university writing has tended, since the 1960s, to avoid discursivising judgements of taste, value, and professional adequacy.

In Deleuze's *La Philosophie Critique de Kant*, already published in French in 1967 – hence before his Spinoza – Deleuze observed, of time, that as long as time remains subordinate to 'the cardinal points through which the periodical movements that it measures pass', it is:

> Subordinate to movement: it is the measure of movement, interval or number. This was the view of ancient philosophy. But time out of joint signifies the reversal of the movement-time relationship. It is now movement which is subordinate to time. . . . Time is no longer related to the movement which it measures, but movement is related to the time which conditions it. (Deleuze, 1984, vii)

Now, we might need to leave it to expert spectators to signal whether or not the performance event – including durational performance – can seem to have 'unhinge[d] time', *even while an older and more stablised time continues to condition its being as performance event.* My 'own sense' (and I use the term advisedly) is that in certain events spectators may seem on occasion to be caught up at the interface between times *sensed*, times *intuited*, times *measured, and* times *recalled.*

Words, words . . .

There is something odd about writing, and associated activities – not least, perhaps, if you are a professional performance-maker. There is nothing natural at all about writing, even if its supremacy in the university context is largely unchallenged; in terms of the digital economy, and in particular of e-mailing and texting, certain widely accessible registers of writing have quietly achieved a renewed currency. Let us not suppose, however, that writing's relations with the world of substance are straightforward. Certain uses of nouns and verbs have been described as 'ontologising', in the sense that they enable writers to *assert being*, while believing that they are merely describing what already exists.

The term 'the body' has been widely used by dance and visual art writers over recent decades, as though its use were unproblematic. Its use itself seems to conjure up something common sensical; yet numerous writers, Hayles (1999) amongst them, have pointed out, *in writing*, that that particular term 'is a congealed metaphor, whose constraints and possibilities have been formed by [a] . . . history that intelligent machines do not share' (p. 284). The term 'the body', when it is used in dance writing (at least with regard to dance expertise), is a nonsense. Those who use it unquestioningly impose an anonymising and de-professionalising label on dancers chosen quite specifically because of their expertise and potential. Second, the term 'the body' itself is a shifter, able to move endlessly between contexts of use, slipping easily, and apparently without loss of identity, into multiple relational sets.

Meanwhile, '*a* body', in Deleuze's indefinite use of the noun, is 'not defined by a form or by functions' – such as might be identified at key stages of professional production. On this basis, 'the performer' – a functional/actional, as well as a professional term, plainly cannot be equated with Deleuze's 'Spinozan' reconfiguring. Yet his notion of a 'living individuality', viewed as 'a complex relation between differential velocities', might permit us to begin to account for some of the ways the professional performer is perceived, by a choreographer. What is ill-assorted with this sort of perspective is not in fact 'the performer' – whom one might view as 'becoming choreographic' – but rather the term 'the/a body', and the processes of nominalisation which produce it, as though it were viewed from a safe distance, and objectified.

On the other hand, as many writers have noted, it is Deleuze's 'second [Spinozan] proposition' – referring 'us to the capacity for affecting and being affected' – which is of particular interest in the context of expert performances, and of what they seem to bring together in relational terms. It is 'affecting and being affected' which seems to challenge certain in(ter)ventions specific to the digital economy. I am reminded here of Ulmer's observation (1994) with regard to affect and 'electronic rhetoric', that what might be at stake in the latter's development is the means to trigger affective memory in users of electronic media. How might a spectator-user (as distinct from a practitioner) be enabled to feel something (qualitatively *more*) *in the event*, in contrast with merely perceiving something?

Let us come back at this point, with that 'qualitatively more' in mind, to the noun 'body' as it is used in performance writing, and thereby to nominalisation – a process of *naming* that tends to confer stabilised being, whereas our concern might actually be with *motion-in-time* (becoming). The noun 'body' is small, hard, and neat, whereas worlds themselves unfold from uses of it. Writing is an economical as well as an ancient symbolic. 'Body', in etymological terms, seems to have been understood to be a container: 'bodig', before 1297, as far as *written* records go, was a trunk or chest (of a man or animal); but originally understood to mean a cask (www.OED.com/). Not solid then, and not simply a matter of 'what a body can do' (i.e. an actional definition), but also something which holds *something different* within it. How curious.

'Body', in *written* record, has been separated from 'soul', since 1240 (and 'soul' will probably have shadowed many of its ongoing uses).

'Body' meaning 'corpse' in written record dates from 1280, bringing with it the opposition 'live' vs 'dead'; 'heavenly body' dates from around 1380, and 'body politic' – 'the nation, the state' – was recorded, in writing, around 1532. It follows that Spinoza's actional approach in his *Ethics* – 'We do not know what *the* body *can do*' (my emphasis) – dating from around 1661–1675, and written in Latin, may well have been informed by at least *this range* of earlier uses. The term is a carrier. It is historically relational; it signals a semiotic domain, rather than that simpler 'meaning' others might be tempted to attribute.

Uses of the term 'the body', in terms of contemporary performance writing, bring with them a veritable network of values *(measures)* and potential unfoldings; its use in that particular context identifies a momentarily and artificially stilled site, within networks of relations, and not a material being as such. In this precise context, whatever the performance-interested academic writer might claim, 'the body' tends to be othered; it is a matter and a measure of expertise – which spectators readily recognise, but may lack the means to analyse as such. It is a matter of signature (the dancer's name signals a complex measure of expertise, artistry, and the potential for qualitative transformation). Its use is a matter, *in the event*, of expert-becoming. 'Expert-becoming' signals the fundamental performance focus on futurity and becoming, which drives the expert practitioner. In this sense, too, in these highly specific situations, unqualified use of the term 'the body' is a nonsense.

A Bergsonian 'becoming'?

The word 'becoming', as most readers of the present volume will be aware of, seems to have retained its currency, across the decades, as well as its qualifier, 'Deleuzian'. It might appear, on that basis, that becoming is able to transcend (historically recorded) time. The notion of becoming certainly seems consistently to appeal to arts-practitioners and postgraduate students in arts-related fields. Yet the term, despite the fact that it is widely linked to Deleuze's writing (in translation, at least), and might then appear, to many English-language readers, to be of the 1980s, was actually borrowed by Deleuze from the late-nineteenth-century to early-twentieth-century philosopher, Henri Bergson (1859–1941). This observation brings us back to historical time, and to what needs to be calculated in its terms.

Bergson's *le devenir* ('becoming') was plainly 'pre-Deleuzian', but it is equally pre-'digital' and pre-'posthuman', *in technological terms*, and in terms of what Rabinow, after Heidegger, calls 'equipment' (Rabinow, 2003), associated with particular technological states. I am interpreting 'equipment', here – because I am a writer – in terms of practical knowledge and its 'apparatuses'; in terms of what their usage teaches; in terms of what users thereby internalise (in terms of expertise), and the actional potential which they also suggest. I am inclined to suppose, in addition, that widespread and repeated uses of 'equipment' are impressed, in terms of actional potential, 'somewhere in the brain' of users. To take up Jameson's (1991) borrowed term, 'cognitive mapping', I would add that technological change, and repeated access to the 'apparatuses' and 'equipment' linked to it, is likely to involve a degree of cognitive re-mapping, or re-configuring.

My more general point here relates to the possibility that 'new work', when it includes the work of live performers and the potential offered by the digital and/or virtual (I am borrowing here the sort of distinction made by A. Murphie (2000)), as well as the live spectator relation, may well be recuperated by some of us – myself included – in terms of the cognitive impress of 'older' technologies, and the 'measures' linked to them. I am myself, from this perspective – speaking professionally – *not yet posthuman*, despite Hayles' (1999) magnanimous turn of millennial account. When it comes then to *what I* might understand of performance-making, and what I might do while spectating, I am supposing that while my sympathies and political engagement remain humanist, my hands-on practices, when it comes to equipment/apparatuses, are simultaneously sympathetic to potential *effects enabled by* the electronic, but lacking in 'equipmental' expertise when it comes either to 'causes' or to real operational processes. Now, some might argue that this familiarity with effects but not with causes is no different from the situation which applied when some of us drew, in performance, on the technical expertise of an expert lighting designer. The difference emerges, as far as I can tell, in the sorts of claims which tend to be made for the virtual, *in writing*, and – more tellingly – *after the event of its emergence*, as was the case for Murphie on Stelarc, in the text cited above, and for Massumi (2002) on the same performance-practitioner.

I want to make two brief points about Hayles' observations on the posthuman ('[t]he great dream and promise of information is that it can be free from the material constraints that govern the mortal world' (Minsky, cited in Hayles, 1999: 13)): (1) there is no possible 'freedom from material constraints',[2] in performance-making *and spectating*, even online: a user is *grounded*; (2) the Spinozan notion of 'the body' as a 'composition of speeds and slownesses on a plane of immanence', if we are to take the Deleuzian at its word, was already a matter of virtuality. The 'extract[ion] of human memories from the brain' has always been attempted by artists, but far from seeking to 'import them, intact and unchanged, to computer disks' (Minsky, Cited in Hayles, 1999: 13), those artists have sought to transform them in terms of qualitative measure on a plane of immanence.

A 'becoming body' tends, as far as performance writing is concerned, to be an identification worded *after the event* of spectating – hence it is an effect. Naming itself consolidates the confusion of effects taken for causes. Performance *writing* persists in reproducing the materially grounded constraints applying to writing, even on-line, at precisely that moment when a writer thematises the virtual. Even where 'posthuman' sympathies are evidenced, the *orders* of expert writing win out, by maintaining certain sorts of spectator perceptions (such as those of the very perceptive Massumi, in relation to Stelarc) and objectifications, through recourse to long-established and conventional registers; and expert writing appeals widely, on that same basis, at least to those of us who work, by and large, within the university community.

The Bergsonian *'becoming'* as Borradori (2000) has pointed out, may be unavoidably 'understood', *by readers* in the later twentieth/twenty-first centuries, *via Deleuze* – hence through and within the play of writing; whereas a *performance* becoming-body, operating outside of writing, is an effect of concrete performance choices, already stablised algorithmically at source. According to Borradori, 'Bergson offer[ed] an authentic conception of difference', which had a particular appeal in post-Second World War Western Europe. 'Authentic' difference, in Bergson, is natural rather than oppositional, and operates through nuance. Yet, whereas nuance can be *effected*, for spectators, its causation is algorithmic, involving concrete choices taken, and is stabilised. If nuances seem – to spectators – to mark 'phase transitions', and to signal 'tendencies', rather more than they are based on clear-cut distinctions, gaps, interruptions, or disruptions, it remains the case that their causes tend to be neatly quantified.

If 'phase transitions', in DeLanda, are 'events which take place at critical values of some parameter . . . switching a physical system from one state to another', so that a 'broken symmetry' emerges between the two states (2002: 18), we might nonetheless need to make a distinction here between performance-making causes and spectator-perceived effects. According to Borradori (2000), what is central here are the ways 'temporality affects the notion of substance'; yet my own argument is that different temporalities affect substances differently, in performance-making and in spectating. In both, perhaps,

> We construct our experience of the world according to two distinct temporal sensibilities . . . because the mind spontaneously 'tends' to process data in two different and mutually irreducible ways, i.e., by way of perception and memory. Memory and perception are thus the two fundamental [time-specific] tendencies underlying our experience.

Deleuze, Borradori notes, 'insists that memory and perception are not available in isolation from one another but are functionally interdependent, which means that memory and perception', for a philosopher, at least, 'do not produce independent kinds of experience'. Instead, 'experience is [an indivisible] "mixture" of memory and perception'. *Outside of writing*, however, in performance-making and spectating, the emphasis shifts and clear distinctions need to be observed. How does an expert performance-maker measure this mix of 'temporal sensibilities' – *for a spectator*? With what effects/affects – where that spectator is our focus? Borradori's response is curiously clear: we 'assess their [operations]' *through intuition*.

'Professional' or 'expert' intuitions

How might we understand 'intuition' where decisions are taken in terms of changing technologies? I need at this point to make a 'knowledge-political', discipline-specific observation with regard to the operations of intuition: because my interest here is performance-making (*outside* – so to speak – the university), I prefer to advocate use hereafter of the term 'expert or professional intuition', or 'disciplinary intuition'. My argument once again is that 'performance-disciplinary intuition' and 'performance expert intuitions' are significantly different from both 'everyday intuition' and 'philosophical intuition'. The primary difference is this: expert/professional intuitions *are constitutive parts of professional expertise*, activated at particular stages *in actual performance-making* – wherein they tend to serve professionals as tools informing decision-making. This expert, performance-disciplinary intuition, in turn, differs from Bergson's 'philosophic intuition'. According to E. Grosz (2000), on Deleuze/Bergson,

> [A]n intuition is a remarkably simple 'concept', whose economy and unity is belied by the (philosophical) language which expresses it. More a 'shadow' (Bergson, 1946: 129), a 'swirling of dust' (132) than a concrete and well-formed concept, intuition is an emergent and imprecise movement of simplicity that erupts by negating the old, resisting the temptations of intellect to understand the new in terms of the language and concepts of the old (and thus the durational in terms of the spatial).

Perhaps Grosz's notion that an imprecise movement of simplicity 'erupts by negating the old' is rather more forceful than is useful here. I prefer to argue that expert intuitions, emergent in particular disciplinary fields, allow 'something' (in the making) to 'feel right', on the basis of which 'new possibilities' can be acted upon. This sense of rightness, and of a 'sudden recognition', according to Ulmer (1994, pp. 142–143), emerges on the basis of a 'cross-modal transfer and transposition across emotional sets', in the contexts of making ('new') work. The affective seems to be constitutive to that context – even where intelligent machines are involved. In Bastick (1985) new links are made *affectively*, and new pathways are opened up. These processes are linked to a 'recentring insight', on which basis an expert practitioner, at a particular stage (time) in the making, and confronted by what might, after Heidegger, be called 'equipment breakdown' (1977, in translation), is able to return productively to the making.

By way of contrast with this sudden and 'imprecise movement of simplicity', analysis, 'which science most commonly utilizes as its method', tends to operate to a different logic: analysis 'decomposes an object into what is already known, what an object shares with others, a categorical rather than an individuating mode of knowledge'. The operations of a 'post-Bergsonian' expert intuition, indeed, drawn on systematically by expert performance-makers, entail 'a mode of "sympathy" by which every characteristic of an object (process, quality, etc.) is brought together, none is left out, in a simple and immediate resonance of life's inner duration and the absolute specificity of its objects'. According to Grosz's reading of the Bergson–Deleuze interface, intuition is 'an attuned empiricism that does not reduce its components and parts but expands them to connect this object to the very universe itself'.

If we return to my insistence on the qualifier *professional or expert*, then the operations of intuition as constitutive but *felt* decision-making processes in the production of 'new work' will exploit, according to Grosz, 'two [professional] tendencies which blur into each other': the first operates in terms of a 'fusional continuum that marks differences in nature and differences of degree'. It tends towards the 'inside', into a 'depth beyond practical utility'. The second tendency which also characterises the intuitive 'is a reverse movement', in which this tendency towards interiority 'sees in itself, in the depths of its own self-immersion, the durational flow that also characterizes the very surface of objects in their real relations with each other'.

Towards a conclusion

To what extent does recourse to 'new technologies', in performance-making, permit performance-makers to engage with notions of human 'interiority', 'depth', 'surface', and memory, *in performance-productive terms*? My earlier observation was that although certain terms used in performance writing ('the body', 'becoming') might seem to transcend their context of use, in fact the *technological* apparatuses, the cognitive impress particular to them, their hands-on uses and users, and the understandings they allow differ between different types and positions of users, at particular times and places. My sense, with expert practices and disciplinary intuitive processes in mind, is that it is *in time* – that is, durationally – that 'surface', 'depth', and 'inside' are economically referenced as a relational set, by performance, on behalf of a spectator.

My sense is first that expert practitioners themselves *expert-intuit* these sorts of relations, seeming to recognise something 'new' in them; second, that they proceed to experiment productively with these insights, *with a particular objective and deadline, in mind*. Third – and importantly – that it tends to be on that complex basis, that practitioners proceed then *analytically* to make 'new (signature) work', via the logics of performance production, in terms of the production values through which *measure* operates.

In other words, 'progress' in performance-making seems to depend upon the ability to bring into a productive interface, the expert-intuitive, the signature-specific, and the performance-maker analytical, with the production logics and equipment specific to the state of the discipline. These processes, where the output of expert-intuitions and the transformative logics of performance production intermingle, will tend to lead to the establishment of sets of performative triggers, some of which entail relational sets instantiated in performance schematics; others will operate through the triggering, *in time*, of a number of performance symbolics, in such a way that a spectator, *in the event*, caught up in the times of perception, might well find that her own memories are productively and meaningfully engaged.

Now, interiorities and depths *can only be intuited*, on the basis of triggers articulated in time, since the former, by definition, have no surface manifestation *of their own*. There are ontological implications: I cannot (and nor can performers), in that event, seem to 'have' a psyche, but I can certainly be enabled to intuit its operations, as well as its site ('inside', and 'deep'). I cannot, in the event (and nor can performers, or *mise en scène*), 'have' memories, but I can certainly be enabled, over (performance) time, to sense that my memories are engaged and activated, by what I perceive. The *extent* (as well as the specificity) of that perceiving is time-specific, and performance-time-enabled.

Where that activation occurs, as I suggested above, performance will seem to 'do more' than it does, and also to 'signify more', and in more complex manner. My feeling is that inferential processing (which *time* takes) tends to operate, in the event, or after it, on the basis of spectator-expert intuition ('Aha!'). Typically, I may well *not* intuit the importance of psyche, in contemporary dance, on behalf of the performers, in the event. In certain instances, but not at all in others, I will spend time, after the event, trying to analyse performance in terms of the putative impress of a performance-maker's psyche. How curious.

It is finally worth noting here that spectators will tend to engage productively with performative triggers, in the event, *and after it*, even when some of those triggers remain, for the expert practitioner, a matter of an undiscursivised intuition. In other words, identification and use of certain performative triggers, folded back onto performance by expert spectators, may well surprise practitioners themselves. In many instances, given the temporal economy of performance, many of these triggers operate in terms of schematic or symbolic hypotyposis (Melrose, 2005), whereby *what is present but incomplete* triggers a spectator's vital contribution, to the event, of something (or things) operating below the level of representation. Economical triggering allows performances to seem to *do more than they do*, and this 'more than' – entailing a constitutive 'not-yet-here' – has very clear implications for the digital-performance-making economy.

Notes

1. This 'more than' differs between performance genres, and it actually serves as a measure of disciplinary identity. The dramatico-theatrical, for example, tends to work significantly in terms of faciality, foregrounded over time.
2. Hayles herself acknowledges this.

References

Bastick, T. (1985). *Intuition: How We Think and Act*, New York: John Wiley.

Bergson, H. (1946). *The Creative Mind. An Introduction to Metaphysics*, trans. M. Andison, New York: The Philosophical Library.

Borradori, G. (2000). 'The Temporalization of Difference: Reflections on Deleuze's Interpretation of Bergson', *Continental Philosophy Review*, Vol. 33: 4 (October).

Bourdieu, P. (2005) [1977]. *Outline of a Theory of Practice*, trans. R. Nice, Cambridge University Press.

Butcher, R. and S. Melrose (eds) (2005). *Rosemary Butcher: Choreography, Collisions and Collaborations*, London: Middlesex University Press.

DeLanda, M. (2002). *Intensive Science and Virtual Philosophy*, London and New York: Continuum.

Deleuze, G. (1981) [1988]. *Spinoza: Philosophie Pratique*, Paris: Presses Universitaires de France; revised and expanded, Paris: Les Editions de Minuit; *Spinoza: Practical Philosophy*, trans. R. Hurley, San Francisco: City Lights Books.

Deleuze, G. (1984). *Kant's Critical Philosophy: The Doctrine of the Faculties*, University of Minnesota Press: Minneapolis.

Deleuze, G. (1988). *Bergsonism*, trans. H. Tomlinson and B. Habberjam, New York: Zone Books.

Deleuze, G. and F. Guattari. (1984). *Anti-Oedipus: Capitalism and Schizophrenia*, trans. Robert Hurley, Mark Seem, and Helen R. Lewis, London: Athione.

Deleuze, G. and F. Guattari. (1987). *A Thousand Plateaus: Capitalism and Schizophrenia*, Vol. 2, trans. B. Massumi, Minneapolis: University of Minnesota Press.

Deleuze, G. and F. Guattari. (1994). *What is Philosophy?*, trans. H. Tomlinson and G. Burchill, London and New York: Verso.

Grosz, E. (2000). 'Deleuze's Bergson: Duration, the Virtual and a Politics of the Future', in I. Buchanan and C. Colebrook (eds) *Deleuze and Feminist Theory*, Edinburgh: Edinburgh University Press.

Hayles, K.N. (1999). *How We Became Posthuman: Virtual Bodies in Cybernetics, Literature and Informatics*, Chicago & London: University of Chicago Press.

Heidegger, M. (1977). *The Question Concerning Technology and Other Essays*, trans. W. Lovitt, New York: Harper Torchbooks.

Jameson, F. (1991). *Postmodernism or the Cultural Logic of Late Capitalism*, Durham and London: Duke University Press.

Knorr, Cetina K. (2001). 'Objectual Practice', in T. Schatzki *et al.* (eds) *The Practice Turn in Contemporary Theory*, London and New York: Routledge.

Massumi, B. (2002). *Parables for the Virtual: Movement, Affect, Sensation*, Durham and London: Duke University Press.

Melrose, S. (2005). http://www.sfmelrose.u-net.com/justintuitive.

Murphie, A. (2000). 'The Dusk of the Digital is the Dawn of the Virtual', *Enculturation*, Vol. 3, no.1 (Spring).

Osborne, P. (2001). *Philosophy in Cultural Theory*, London and New York: Routledge.

Rabinow, P. (2003). *Anthropos Today: Reflections on Modern Equipment*, Princeton and Oxford: Princeton University Press.

Ulmer, G. (1994). *Heuretics: The Logic of Invention*, Baltimore and London: Johns Hopkins University Press.

Zizek, S. (2004). *Organs Without Bodies: On Deleuze and Consequences*, London and New York: Routledge.

Web pages consulted

http://www.seop.leeds.ac.uk/index.html, accessed October 2005.

http://www.OED.com, accessed October 2005.

http://www.sfmelrose.u-net.com, accessed October 2005.

2
Truth-Seeker's Allowance: Digitising Artaud

Steve Dixon

Introduction

Antonin Artaud was a truth-seeker, a philosopher of the theatre. Thus, anyone following his philosophies, as I try to through the work of my multimedia theatre company *The Chameleons Group*, must necessarily have faith in and accept the difficult (and some might say discredited) concept of *truth*. A truth-seeker's allowance must be granted to such people, just as a financial 'job-seeker's allowance' is granted to those unemployed in the UK who can demonstrate they are actively seeking work, whether or not they are successful. The truth-seeker's allowance provides a benefit-of-the-doubt concession to those wedded to traditional philosophical, spiritual, or structuralist ideals which espouse that *truths*, universal or otherwise, actually *do* exist. They can be revealed, articulated, and shared. Artaud sought all three outcomes respectively: through theoretical meditations, performance manifestos, and theatre productions.

I am primarily involved in the first and last of these – theory and practice – and prefer for the middle category to follow Artaud's own manifestos and Andre Breton's *Surrealist Manifesto* rather than inventing any of my own. But there is a gap in the manifesto market that will surely be exploited in the future, since to date there have been no seminal manifestos on the conjunction of performance arts and digital media. This chapter may even provide an embryonic rehearsal towards its development: a fumbling early sermon prior to some future evangelical ejaculation about how the transformational capabilities of computers within theatrical contexts can relate to and extend philosophical theories, Artaudian and surrealist conceptions of art, and, most crucially of all, notions of *truth*.

My multimedia theatre practice as director of *The Chameleons Group* fuses digitally manipulated video projections with live theatre to follow Artaudian and surrealist ideas, and to update their very particularised notions of *truth* for a contemporary audience. So part of my allowance entails ready engagement with these theories, and an exploration of pertinent perspectives on how stage and screen conjunctions operate theatrically, but sidesteps and dispenses with the array of established and diverse philosophical theories on *truth* (semantic theories, correspondence theories, pragmatic theories, deflationary theories, etc.). But if we accept that theatre, like philosophy, involves some pursuit of truth, my question is how do new media technologies affect the search, and the specific path of the media-theatre truth-seeker?

Of course, the very idea of artistic truth has become theoretically unpalatable to many following postmodern and deconstructive debates, but such discourses and their influence have recently lost momentum, returning many artists and commentators to more traditional hermeneutic perspectives and philosophical enquiries. It is from that position that I write, and produce theatre. Paradoxically (and in analytic philosophy the paradox famously proves the flaw in the argument and mitigates against its *truth*), the form and style of *The Chameleons Group*'s theatre owes much to the artistic conventions of postmodernism and deconstruction. But the underlying philosophies and intentions behind it are avowedly modernist – we are interested in using multimedia in live performance to explore and expose serious existential issues, altered mental states, and metaphysical notions. These aims we share with the surrealists of the early twentieth century, and with Artaud, a 'lapsed' surrealist who was the Director of Surrealist Investigations in 1925, but later dissociated himself from the movement following Breton's political alignment of surrealism with communism.

Artaud was a theatrical philosopher, a truth-seeker, and it is somewhat ironic that one of the leading commentators on Artaud's work, Jacques Derrida, is also arguably the twentieth century's most influential challenger and destabiliser of the very notion of *truth*, at least in any absolute form. Derrida's deconstruction of language parallels Artaud's celebration of theatre as a plague. The plague analogy highlights a subversion of morality and society, where barriers become fluid, order disappears, and anarchy prevails (Artaud, 1974 [1938]: 7). Derrida celebrates the same collapse of boundaries, the same scourge and breakdown of order and signification, as language itself becomes a plague, a cellular virus, splitting and infecting its very self. But whilst Derrida's linguistic plague darkens, divides, and undermines notions of meaning and truth, by contrast Artaud's plague is bright and blinding in its revelations, seeking transcendent, unifying meanings, and nowadays highly unfashionable universal *truths*:

> If fundamental theatre is like the plague, this is not because it is contagious, but because like a plague it is a revelation, urging forward the exteriorisation of a latent undercurrent of cruelty through which all the perversity of which the mind is capable, whether in a person or a nation, becomes localised. (Artaud, 1974 [1938]: 19)

Derrida took Artaud's theories to brain rather than to heart, and Derrida's championing of the theatre of cruelty is predicated upon a belief in Artaud's drive to destroy theatrical language and metaphor – on its de(con)struction rather than its creation. But his position on theatre *per se* is antagonistic: 'The theater itself is shaped and undermined by the profound evil of representation. It is that corruption itself. Theatrical representation ... is contaminated by supplementary re-presentation' (Derrida, 1976 [1967]: 304). His distaste for the theatrical continues in his analysis of the actor:

> Vice is his natural bent. It is normal that he who has taken up representation as a profession should have a taste for external and artificial signifiers, and for the perverse use of signs. Luxury, fine clothes, and dissipation are not signifiers incidentally coming about here and there, they are the crimes of the signifier and the representer itself. (Derrida, 1976 [1967]: 305)

Derrida's poststructuralist thought is descended from Nietzsche, and extends his notions of the world as a constant flux, where all beings and entities are unstable and in a state of becoming. Theatre and performance's process (writing, preparation, rehearsal) adheres to this philosophy. But in its manifestation – that is to say, in its essential ontology – it seeks just the opposite: to create and stabilise a world, to unite time and space, to *become*. So whatever his brilliance and influence, in relation to theatre, let us put Derrida and his followers firmly to one side as part of the concessionary 'allowance', or we may never get anywhere in our discussion of *truth*.

Practicing multimedia *truth*-theatre

Artaud's *truths* were many and varied, and sometimes contradictory, and centred on ideas such as the double, cruelty, and the body. Taking these broad themes and expressing them in multimedia theatre form is in itself no great challenge; in fact it is quite simple (perhaps *truths* were ever thus). For example, in *Chameleons 4: The Doors of Serenity* (2002) our live performers constantly confronted their projected digital doppelgängers (the double), screen images included characters being whipped with bloody entrails and others self-mutilating (cruelty), and 'the body', whether naked or clothed, triumphant or abject, always dominated both stage and screen (we use no props or set, only a screen, on which changing images of the bodies of the performers are omnipresent). But what types of *truth* or *moments of truth* these media-theatre images expressed and revealed is quite another matter. For audience members we spoke to, certain moments and images were considered powerful and disorientating; some spoke of bravery and honesty, but few mentioned *truth*. As I recall, neither did we the performers; the notion of *truth* was not fundamental at the time and I must confess it has only now occurred to me, several years later. But the process of devising the work involved psychological risk and personal pain as we tried to excavate and exorcise inner fears, fantasies, and 'demons', and thus truthfulness, if not *truth* itself, was undoubtedly invoked.

Artaud tells us that 'acting is a delirium like the plague', and his vision of the actor in delirium, the martyr burned alive and still signalling through the flames, remains one of the most potent and spiritual articulations of acting theory. For Western performers, moments that even come close to this grand metaphor are rare. Our use of the actor on stage and the double on screen, the performers acting with and against themselves at worst offers the actor two bites of the cherry, and at best opens the possibility of some synergetic alchemy which might approach this notion of flame-licked delirium. Some *Chameleons 4* images even directly reference Artaud's metaphor – a light-hearted live stage scene where the two male characters, a cyborg and a devil, first meet is played in front of a projection of the cyborg character, a noose around his neck, screaming in agony in the midst of lapping, digital hell-fire whilst the devil, in miniature, dances on his shoulder, and the ghosts of the two women characters wander, lost and blindfolded, in a darkened, background purgatory.

It is no accident that the live scene is played 'light' and the projection sequence is distinctly 'heavy', and this provides a key to understanding

Figure 2 In *Chameleons 4: The Doors of Serenity*, the fusion of digitally manipulated video imagery with live performance conjures a form of 'synergetic alchemy' to explore and evoke Artaudian conceptions of the double, cruelty, and the body. Clockwise from top left: Steve Dixon, Wendy Reed, Anna Fenemore, and Barry Woods

the potent praxis of new media-theatre. Not only does the simple coun-
terpoint and contrast between the stage and screen texts 'work' theatrically; not only does the composite help to disorient the audience's senses
in order to take them into Artaudian realms; but the fundamental *truth* is
that such an effect and image cannot be achieved by live theatre alone.
The use of technology in creating the projection enables each actor
(in video recording) to reach a point of acting *truth* through focussed
preparation and multiple takes where they concentrate on only one
outcome – a moment of horror – rather than the many and varied ones
required in the course of a live performance; and (in post-production)
technology enables a complex montage layering of four separate, individually honed performances to be conjoined and incorporated within
an animated graphical environment (hell and its fires). The projection,
the visual intensity, and acting demands of which might be conceived
as a climactic ending were it performed on stage is placed with ease
within the expositional first act, and is juxtaposed against a prosaic live
prologue. It is a parallel spatial and psychic dimension: the characters
are revealed to inhabit and haunt both their own inner (screen) and
outer (stage) spaces and psyches.

One of my points here, in regard to recorded media's relative
empowerment of acting *truth* in comparison to live performance,
relates back to one of the earliest attempts to define and differentiate
the separate ontologies of film and theatre, Allardyce Nicoll's *Film
and Theatre* (1936), which places the notion of *truth* at its core. Nicoll
maintains that audience orientations when experiencing the two forms
are quite different since theatre has an inherent falsity acknowledged
by all – 'dramatic illusion is never ... the illusion of reality; it is
always imaginative illusion' (1936: 166) whereas film purports to
truth, and despite the fact that the idea that 'the camera cannot lie'
has been disproved, 'in our heart of hearts we credit the truth of the
statement' (p. 167). He maintains that the individualisation process
in film performance enables and indeed demands greater complexity
in characterisation, and that: 'What we have witnessed on the screen
becomes the "real" for us' (p. 171), an idea that anticipates postmodern
media theories as exemplified by Jean Baudrillard. Nicoll's ideas are
interesting in contrasting film as 'truthful', 'complex', and 'real' with
the falsity, simplicity, and illusion of theatre. Whether or not one is
inclined to agree with his analysis, I would nonetheless maintain that
his thesis still holds 70 years later within multimedia theatre, where
the screen space(s) and their images generally tend to explore notions
of reality and *truth* far more than the live performers within the stage

space attempt; and where Nicolls' notions of filmic 'complexities' are doubly extended through digital manipulations and metamorphoses.

Truth and confrontation

In the *Surrealist Manifesto*, Breton discusses the language of surrealism as a dialogue in which two thoughts confront one another, the one reacting to the other. He describes how the one thought or idea will treat the opposing idea as an enemy. Through this confrontation the original idea will become distorted, will change its very nature. This metamorphic tension between two signifiers – in our case, the stage and the screen – is in constant play during our performances, and in many ways defines the essential ontology of multimedia theatre. In Chameleons Group performances, the screen and stage embodiments are in constant battle with each other, like schizophrenic personalities. But it should be noted that this central surrealist conception of an opposition and enmity between two ideas or forms has itself been a central pillar of criticism by those opposed to theatre's incorporation of media projections. Opponents fiercely contest that there is a mismatch of media and a corruption of theatre's purity as a live form: media projections do not enhance the intellectual power or visual spectacle of theatre, rather their technological intrusion is alien; the two forms are aesthetic enemies. In 1966, Susan Sontag summarised the debate in relation to the use of film projection in theatre:

> The big question is whether there is an unbridgeable division, even opposition, between the two arts. Is there something genuinely 'theatrical', different in kind from what is genuinely 'cinematic'? Almost all opinion holds that there is. A commonplace of discussion has it that film and theatre are distinct and even antithetical arts, each giving rise to its own standards of judgement and canons of form. (1966: 24)

The wider tension between theatre and technology, of course, goes back much further. In *The Poetics*, Aristotle placed spectacle firmly at the bottom of his list of constitutive elements of dramatic tragedy, and Jacobean playwright Ben Jonson fought famously and furiously against the spectacular designs of Inigo Jones that threatened to upstage his text. The same argument raged in the early twentieth century when directors such as Erwin Piscator incorporated film footage into theatre perform-ances. His production of Ernst Toller's *Hoppla Wir Leben!* (1927) received equal acclaim and condemnation, including from Toller himself, who two years later in 1929 described Piscator's attempt to unify theatre and film as 'a mistake', since he considered that 'the two arts followed different laws' (Hern, 1972: 85). But in the same year, American theatre designer Robert Edmund Jones published a contribution to the *Encyclo-paedia Britannica* entitled 'Theory of Modern Production', which argued that the fusion of theatre and cinema prompted a unique and potent new artform: 'In the simultaneous use of the living actor and the talking picture... there lies a wholly new theatrical art, an art whose possib-ilities are as infinite as those of speech itself' (Jones, 1929: 40). He argued that film offered a resolution of the theatre dramatists' problem of how to effectively express the inward reality and subconscious of their characters, since film offered 'a direct expression of thought before thought becomes articulate... the moving picture is thought made visible' (Jones, 1929: 40). In his 'theatre of the future' the live actor would thus represent the character's outer self and the screen imagery the inner world of imagination, subconscious, and dream: 'the two worlds that together make up the world we live in' (Jones, 1929: 40). The synthesis of film and theatre became an obsessive theme that Jones developed in his book *The Dramatic Imagination* (1941), the leading textbook for American theatre students at the time, and which he later reworked and distilled as great orations during years of lecture tours in the 1940s and 1950s, discussing in quasi-Freudian terms the relationships between stage and screen; and between existential and theatrical *truths*:

> At the root of all living is a consciousness of our essential duality... and now there is a way to say all this in the theatre – simply, easily, straight and plain. ... On the stage: their [character's] outer life; on the screen: their inner life. The stage used objectively, the screen used subjectively, in a kind of dramatic counterpoint. Not motive as it is revealed in action, but action *and motive* simultan-eously revealed to us. The simultaneous expression of the two sides of our nature is an exact parallel to our life process. We are living in two worlds at the same time – an outer world of actuality and an inner world of vision. (Jones, 1992: 77)

There are clear relationships here with both surrealist and Artaudian understandings of the nature and ultimate meaning of art, and with art's potential to use clashes, conflicts, and 'doubles' to express universal forces and inner worlds; and alternate realities that go far closer to notions of *truth* than quotidian realities. In the *Surrealist Manifesto* (1924), Breton positions surrealism ideologically as 'a belief in the superior reality of certain forms of association hitherto neglected, in the omnipotence of the dream, in the disinterested play of thought', and this premise has consciously or unconsciously become a central feature of much experimental multimedia theatre including my own, where the enactment of dreams and dreamscapes are pivotal, and the dualities and conflicts of combining media projection and live performance are channelled and celebrated.

In *Chameleons 4*, the interrelationships between live and virtual are used to help awaken and ignite another Artaudian concept, 'physical hieroglyphs'. Influenced by the mudras and physical signification systems of Balinese dance, Artaud believed the use of symbolic or abstracted gestures and physical tableaux was a route to a new theatrical form that was both poetic and occult. This idea is developed throughout the *Chameleons 4* performance, beginning with a scene where the two female characters, a vampire bimbo and a genital-less *femme fatale*, approach the audience like smiling air stewardesses and repeat a series of complex arm and head gestures, whilst the background screen runs multiple composite images of them doing the same in wide-shot and close-up. As the sequence progresses, the gestures and 'hieroglyphs' metamorphose to become increasingly convulsive and grotesque, slowing down to a painful, spasmodic minimalism on stage as the women's smiles become grimaces, whilst the speed and abandon of their screen counterparts' gestural dances intensifies. Later, these kinetic hieroglyphs mutate and reappear in numerous forms, notably a ritual staged live between the vampire and the devil where each use perverse sadomasochistic and quasi-religious gestures including rapid genuflections with their fists, where they violently slap and beat their own faces, hips, and genitals. As this takes place, the projection image depicts the devil slowly convulsing in response to the movements of the face and body of the vampire whose quarter size figure appears to be underneath his skin, inhabiting him, and stretching his epidermis as she appears to flow around his body, like blood. The stage-screen sequence is disquieting yet beautiful, and was inspired by the final line of Andre Breton's *Nadja* (1960) [1927] about his lover who sank into madness: 'Beauty will be CONVULSIVE or it shall not be.'

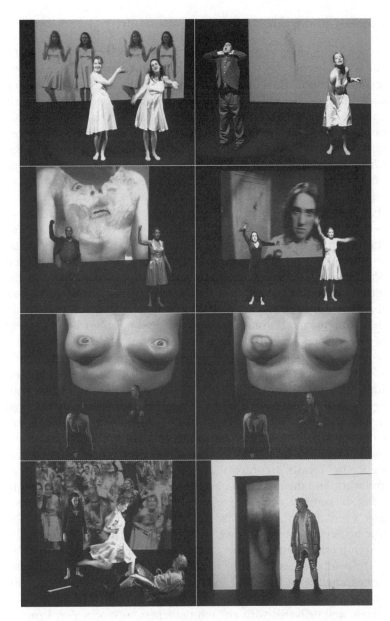

Figure 3 The Chameleons Group utilise the stage and screen spaces to conduct 'doubled' experiments with Artaud's notions of 'physical hieroglyphs' (top 4 photos) and surrealist conceptions of comedy and insults (bottom 4 photos) in *Chameleons 4: The Doors of Serenity*

Comedy and truth

In describing certain sequences from *Chameleons 4*, I have not so far conveyed a complete and *true* sense of the performance overall. The scenes described relate to only the darker aspects of Artaudian theatre, containing elements such as those he once summarised as the imagistic content of his film scenarios: 'eroticism, cruelty, the taste for blood, obsession with the horrible, dissolution of moral values, social hypocrisy, lies, false witness, sadism, perversity' (Artaud, 1930: 23). But *Chameleons 4* is also very much a comedy, since humour serves as one of the most direct routes to audience response and we are concerned to engage and draw them in rather than alienate them, in the same way that director Richard Foreman has explained that his 'goal has always been to transcend very "painful" material with the dance of manic theatricality' (1985: 131). One does not immediately associate Artaud with comedy, yet a close reading of plays such as *The Spurt of Blood* and *There is No More Firmament* reveals Artaud's surrealistic delight in the ultimately comedic absurdities of human existence. In his analysis of the Marx Brothers' movies, Artaud also propounds radical theories on the disruptive and transformative potentials of what he calls 'the power of physical and anarchic dissociation in laughter'. In one *Chameleons 4* sequence, the *femme fatale* accuses the devil for inappropriately staring at her breasts. As he protests his innocence, she walks around him on stage shouting angrily, and finally kneels down in front of him with her back to the audience. On the screen, her movements are mirrored in a reverse, mid-shot of her face and torso, facing the audience. 'Is this what you want to see', she says as she simultaneously pulls down the front of her dress on stage, where we see her back, and on screen where we see her breasts. Her nipples are eyes, which stare back at the devil and wink, and then metamorphose into two female mouths that begin a fast, bickering argument, one angrily denouncing the invasive male gaze whilst the other welcomes the attention her breast attracts. The row between the high-pitched, speeded-up mouths and voices finally descends into a comic argument about whose breast is larger and firmer.

The conflict enacted here between the two sister characters continues throughout the performance, and the use of insults between them, both on stage and screen, intensifies throughout. Their first exchange is on screen, and delivered with an arch politeness, as if at a vicar's garden party: 'My sister has no pussy, but she still reeks of piss' says the vampire bimbo; 'My sister has kipper flaps, because she is a slag', retorts the genital-less *femme fatale*. The sexually related insults continue until a final exchange where, following the vampire's taunt about her sister being 'a toxic laminated half-woman', the *femme fatale* screams: 'you bag of blood sap, who leaks out like a dog on heat, you're a blonde idiot bitch whore who shags donkeys. Why don't you shove your head up your porcelain prickly crab lice flange, you fucking shit fuck slut.' Our group is attracted to the use of insults since in an Artaudian sense it is a unique and specific form of cruelty and violence, and a form that is socially acceptable as it is verbal and emotional rather than physical. Insults are also peculiarly ambivalent entities, since the same insult can be affectionate and funny, or vicious and malevolent, depending on the intention and the way it is delivered. For the surrealists, as Luis Bunuel put it, the insult was a forceful means 'to explode the social order . . . We exalted passion, mystification, black humour, the insult and the call to the abyss' (1984: 107).

Conclusion

I began by noting that my multimedia theatre practice seeks to follow Artaudian and surrealist ideas, and to update their very particularised notions of *truth* for a contemporary audience. That contemporaneity must acknowledge the *truth* that underlies comedy, the expletive insult, or some men's guilt at their breast-staring compulsions, as much as the *truth* that may be enshrined in some sublime aesthetic image. But this is not to offer evasive platitudes about plenitudes of truth to shirk from my task of truth-seeking.

Artaud's vision of *truth* was a theatre of cruelty 'where violent physical images pulverise, mesmerise the audience's sensibilities, caught in the drama as is in a vortex of higher forces' (1974 [1938]: 63), and it is this vision that *The Chameleons Group* attempts to embrace and update for the digital age. The use of digital projections of the live performers' 'doubles' invokes a different 'theatre and its double' than Artaud envisaged, though one that adheres to his original poetic and occult premise. In his discussion of theatre as alchemy in the 1930s, Artaud became the first person to coin the term 'virtual reality'. For him, since alchemical signs are 'like a mental Double of an act effective on the level of real matter alone, theatre ought to be considered as the Double, not of this immediate, everyday reality . . . but another, deadlier archetypal reality' (ibid., 34). A doubled reality or 'virtuality' became the fulcrum for Artaud's theatre of cruelty, and that core of virtuality continues as a primary theme but in a different form for The Chameleons Group.

Nietzsche, who spouted great aphorisms, was also deeply sceptical about the notion of *truth*. 'Truth', he said, is nothing but 'a mobile army of metaphors' (1995 [1895]: 1248). His philosophy attempted to destabilise and undercut Platonic dualism, a 'two-world' view that he also saw as central to modern science in its distinction between observable phenomena and underlying forces. For Nietzsche, the distinctions between 'appearance' and 'reality' expounded by both Platonism and science constituted a false and hardened 'mythology' of dualisms (reason and irrationalism, heaven and earth, mind and body, sensation and thought), whereas in his view there was only one world, the world we find around us. Today, that world (or at least, the Western one) is filled with new technologies, and their incorporation into theatrical practice brings into being the potential for a synthesis where the two elements – stage and screen – are not Platonic or scientific dualisms, but make up a new 'one world' of conjoined 'mobile metaphors' capable of expressing nothing more, and nothing less, than *truth*.

References

Artaud, Antonin. (1930). *La Nouvelle Revue Francaise* (June).

Artaud, Antonin. (1974) [1938]. *The Theatre and Its Double. Collected Works*, Vol. 4, London: Calder Publications.

Breton, Andre. (1924). *Surrealist Manifesto*. http://www.seaboarcreations.com/sindex/manifestbreton.htm, accessed January 2006.

Breton, Andre. (1960) [1927]. *Nadja*, New York: Grove Press.

Bunuel, Luis. (1984). *My Last Breath*. Trans. Abigail Israel. London: Vintage.

Derrida, Jacques. (1976) [1967]. *Of Grammatology*. Baltimore, Maryland: The John Hopkins University Press.

Foreman, Richard. (1985). *Reverberation Machines: The Later Plays and Essays*. New York: Station Hill Press.

Hern, Nicholas. (1972). 'The Theatre of Ernst Toller'. *Theatre Quarterly*. Vol. 2, no. 5: 72–92.

Jones, Robert Edmond. (1929). 'Theory of Modern Production'. *Encyclopaedia Britannica*, 14th edition. UK: Cambridge University Press.

Jones, Robert Edmond. (1992). *Towards a New Theatre: The Lectures of Robert Edmond Jone*. Transcribed and edited by Delbert Unruh. New York: Limelight Editions [Lectures first delivered 1941–1952].

Nicoll, Allardyce. (1936). *Film and Theatre*. New York: Thomas Y. Crowell Company.

Nietzsche, Friedrich. (1995) [1895]. 'Twilight of the Idols: Or, How to Philosophize with the Hammer'. Trans. Richard Polt, in Steven M. Cahn (ed.) *Classics of Western Philosophy*. Cambridge, Indianapolis: Hackett Publishing Company.

Sontag, Susan. (1966). 'Film and Theatre'. *TDR: Tulane Drama Review*. Vol. 2, no. 1 (T33): 24–37.

3

Transformed Landscapes: The Choreographic Displacement of Location and Locomotion in Film

John Cook

Although choreographic and cinematic collaboration began over 110 years ago, little has been written on the transformative kinetic interplay between the two. In this essay I examine works by three choreographers and filmmakers, Swiss director Pascal Magnin's *Reines d'un Jour* (1996), Flemish choreographer and director Wim Vandekeybus and Ultima Vez's *Roseland* (1990), and Canadian director Bernar Hébert and Canadian choreographer Édouard Lock and La La La Human Steps' *Velazquez's Little Museum* (1994), exploring the ways film and video make it possible for the choreographer to displace, without replacing, the presence of the performer. Thus, how can the collaboration between the filmmaker and choreographer move the performers and viewers to a new place/si(gh)te while making the 'place' move at the same time? The choreographic language speaks to and with the physical location, a location that now seems more real or imbued with more 'depth', than the theatrical setting of a theatre. In doing so, the movement possibilities and the performers', as well as the viewers', somatic experiences expand into a new realm of techno-presence.

A variety of possibilities exist for techno-transformation of the dancing body. It is not my intention to address all of these approaches here. Rather my focus centres on film and video and not on computer-based technology; nor do I examine technology as an aspect of live performance, rather my narrow interest is on dances that are made specifically for film and video as independent performance productions.

As a further disclaimer, I recognise the significance of gender, ethnicity, race and sexuality in the dancing bodies I will be writing about. However, my first concern is to argue for the transformative role of technology in kinetic expression. This expression remains constrained in the films and videos I discuss by gender, race, sexuality and ethnicity – though only the latter two are consciously articulated in any of the films. Attempts to res(igh)te bodies outside of these parameters require another argument.

Dance provided the early cinema with a ready and appealing subject – a subject that shared the 'the ability to represent motion and temporal duration' (Doane, 2002: 24). With short routines, numerous solo artists, no need for spoken or written text and at times scantly clad women, the moving images of dance proved to be ideal for the new moving pictures.

While dance continued to play an important role in cinema after the first decade of the twentieth century, it rarely functioned as the sole subject matter for films. Maya Deren's *Study in Choreography* (1945) marks a pivotal shift in the use of dance as film and film as dance. Her four-minute piece, without sound or colour, defines the director's absolute position on dance film as noted by Mark Franko in *Aesthetic Agencies in Flux* as 'a dance so related to the camera and cutting that it cannot be performed as a unit anywhere but in this particular film' (Franko, 2001: 140).

Deren's stringent definition of dance film certainly helps to clarify a complex genre; which easily applies to the three films *Reines d'un Jour* (hereafter *Reines*), *Roseland* and *Velazquez's Little Museum* (hereafter *Velazquez*), under discussion here. This is not to suggest that the films cannot serve multiple functions. Mary Doane in her study of cinematic time recognises as much in her investigation of the presentation and reception of early cinema. 'Yet it is important to emphasise that notions of film as record and film as performance/display are not necessarily contradictory or incompatible' (Doane, 2002: 24).

An excellent example of a film that is both a documentary of a dance and dance film is Chantal Akerman's documentary of Pina Bausch, *Un jour Pina m'a demandé*,[1] which not only documents both the creative process and the performance, but also imitates the dance theatre praxis in its filming and editing. For me a more inclusive definition of dance film might read – a film that depends on dance and a choreographic structure while employing the camera and editing to further the kinetic understanding of performed movement. Or as Sherril Dodds argues in her book *Dance on Screen* 'that video dance derives from the triadic relationship between the motion of the physical body, the camera and the cut. Thus definitions of "dance" and the critical apparatus which surrounds it are challenged and displaced' (Dodds, 2001: 170–1).

I have chosen these three films in part because they all fit under the rubric of physical theatre, described below. Further, each employs what I have defined as filmic kineticism, discussed below, in the making of the films/videos.[2] Finally, each film demonstrates a clear collaboration between the director(s) and choreographer(s) in creating the final product.

Filmic kineticism envisions the camera as an instigator and participant in the kinetic fulfilment of the choreographic design. The camera functions not only to capture or record images, but also to increase the kinesthetic exchange between audience and performer. In doing this, the camera performs a double role, one as an extension of the spectators visual sense, and secondly, as an apparatus that engages their tactile sense.

In a reaction against the formalist aesthetics of Cunningham and the non-narrative semi-pedestrian works of many of the postmodern choreographers, a number of non-American choreographers reintroduced 'narrative', in part reclaiming the expressive technique of the modern dancers of the 1930s and 1940s, but with a sharper political and social agenda. In the visual arts a similar reaction came from artists who rejected the absence of the figure in Abstract Expressionism. According to Mark Franko, the abjection of a 'politico-expressive' in favour of an 'aestheticist perspective' in Cunningham represents a move backwards in the history of dance (Franko, 1995: xii). I would read dance-theatre as exemplified by the creations of La La La Human Steps and Ultima Vez as a leap forwards then, building on Franko's implied 'progresses' of the early moderns, infusing their expressivity with a contemporary, transfigured politicised body – a body transfigured by a somatic reaction to personal violence, an increase in inter-personal risk and the need to forthrightly address sexuality and gender in performance. Such transformations took place not only in regards to content and presentation of material, but also in use of music, costume and set design that influenced and even created kinetic vocabulary.

Cameras possess the potential to transform stage bodies into screen bodies. In some cases, these bodies remain fixed by the two-dimensional format of video or film projections. They may shrink or expand according to the medium or locale of presentation. However, in the case of the films under discussion here, the camera functions in a more complex and integrated role in the filmed/video-taped choreography. In *Roseland* the camera offers views of the dance from a variety of positions unavailable to an audience during the live performance; whereas *Velazquez* and *Reines* transport the audience to a space/location unavailable in live performance.

A privileged access democratically distributed through the medium of television, movie projector or Internet challenges the economic division of differential theatre seating. In all three works, the collaboration between director and choreographer brings the audience into a new relationship with the performers and performance. Further, the performer her/himself revisits the performance now as an audience member able to engage her or his own image, and even, corporeality with the camera. Filmic kineticism extends our engagement by linking the visual and somatic phenomena of performance from both the dancer's and audience's perspectives.

While I would not like to apply Michael Taussig's argument on Mimesis to the dance works under review I do find his discussion of Benjamin to be very useful in furthering my own reading of filmic kineticsim or tactile vision in dance film/video. In his questioning of the visual dominance or ocularcentrism in Euro-American culture, Taussig (1993) proposes that touch aided by vision is more used and useful in navigating architectural and social space than vision alone. Addressing the questions: 'How do we get to know the rooms and hallways of a building? What sort of knowing is this? Is it primarily visual? What sort of vision?' (Taussig, 1993: 26). The author employs Benjamin's argument from *The Work of Art in the Age of Mechanical Reproduction* to counter claims for the superiority of vision in knowledge acquisition.

Applying Taussig interpretation of Benjamin to the extended eye of the audience, the camera, we can connect Benjamin's concept of 'physiological knowledge built from habit' (Taussig, 1993: 26) to the choreographic use of repetition and intimate screening, via zooming, in *Reines, Roseland* and *Velazquez*. Taussig could easily be describing a theoretical framework for the opening scenes of *Reines* or *Roseland* in his statement:

[T]ouch and three-dimensioned space make the eyeball an extension of the moving, sensate body? Which is to say, an indefinable tactility of vision operates here too, and despite the fact that the eye is important to its channelling, this tactility may well be a good deal more important to our knowing spatial configuration in both its physical and social aspects than is vision in some non-tactile meaning of the term. (Taussig, 1993: 26)

Filmic kineticism attempts to bridge the dimensional divide between screen bodies and stage bodies. Consequently, in the work of a few directors, camera operators and editors, they bridge the divide between the visual and the somatic. In these cases a new spatial and sensual performance environment emerges. One that makes intimate the act of choreography and dancing. In no way can this experience replace the live stage event, nor does it try to. Instead, another genre forms made of flesh and tape, of light and sweat. Physical theatre lends itself to such collaborations due to its extreme kinetic language and individual accessibility of performers. Each film bears the unique stamp of director, choreographer, designer and dancer. It is my attempt only to expose moments of collusion and empathy.

Building on my theory of filmic kineticism, this essay expands the scope of the transformative possibilities of technology (the camera, film, video) from an extension of the choreographic vocabulary to a re-inscription of the entire mise-en-scène. The camera enables the choreographer to relocate the dance physically and kinetically in space and time.

Individual cases

Reines d'un Jour (Queen for a Day) (1996)

The first dance film I will discuss is the only one of the selection not based on a stage production. A collaborative project choreographed by six dancers and directed by Pascal Magnin, *Reines d'un Jour*, was filmed in a small village among the Swiss Alps. A hidden narrative unfolds as the choreography flows with the topography of the region.

Not until the very end of the video do we learn the tale on which the dance is based upon. The video opens with three women running up a hill in short dresses clutching their shoes in hand. Three men race up the hill in suits and dress shoes. As one performer falls to roll down the slope others leap over her only to succumb to the same gravitational force. The camera follows the tumbling bodies down, keeping their full figure in frame as long as possible. We see the elongated forms rolling with us, to us and from us as feet and legs scurry over the fallen dancers. Close-up shots articulate the fall and rise of the dancers, actively shadowing the kinetic impulses that initiate movement or transfer it from one performer to the next. Only sounds of scampering feet and exhaled exertion can be heard. The rhythmic structure for this first choreography is provided through editing alone.

Magnin expands the choreographic perception of *Reines* by opening up the breadth and depth of the performance plane. Figures emerge from an alternative time and space onto the screen. By framing the dances with the Alps as a backdrop and isolating various parts of the body, in particular the feet, the camera continually transforms the choreographic sense of scale back and forth from expansive to intimate.

Reines can be broken up into three broad developmental structures. Part one, consisting mostly of running and rolling up and down the side of the meadow, develops a kinetic relationship between the dancers and Nature. As they transition from the solitude of the mountain to the settlement of the village, they pass through a series of ritualised conflicts imitating the cattle vying for position/rank in the herd and playful assertions of mischief with the village children. As the dancers fully integrate into the village community, they participate in a communal gathering, sharing dances and wine.

They acculturate to the mountain, and its community, by trading their street shoes for Swiss hiking boots while succumbing to the incline. They no longer try to defy the slope, but instead throw themselves into the landscape developing a physical theatre movement vocabulary in dialogue with it. Common objects, such as hiking boots and wine bottles, become choreographic tools re-circumscribed by the camera and editing for producing movement phrases.

Cinematic time and natural space replace real time and stage space as the kinetic intrusions by the dancers blend into, then overcome, the rhythm of village life. The inhabitants accept the performers into their daily narrative as if the story being retold, an oral tradition, has become a kinetic tradition existing in a supra-narrative plane.

Magnin achieves what Maya Deren set out to do in her dance film. For Deren the limitations of dance arise from the limitations of architecturally defined space germane to live performance. The mobility of the camera and the manipulations of the editing disrupt such limitations and transfigure them. Through the agency of the camera and editor, 'a whole new set of relationships between the dancer and the space could be developed' (quoted in Franko, 2001: 141). Magnin goes even further, extending beyond 'transfiguring space' to transforming our kinetic experience of the dance through the videography and editing.

Filmic kineticism plays an important role in *Reines*. One frequently used technique involves isolating, through close-ups, first the kinetic impetus then the resulting action of a movement combination. By leaving out many of the transitional movements, the editing replaces the physical transitions with filmic ones. The camera captures intricate foot patterns, falls, embraces and jumps often juxtaposing them against pedestrian or 'natural' movements as reciprocal choreographic interpretations further bridging real space with choreographic space.

Reines comes to an end somewhat abruptly with an elderly woman narrating the tale of the three maidens who provoked God by dancing too much and transgressing against social propriety. In a moonlit lake with floating candles, the three female dancers perform a lyrical dance resulting in their symbolic death. Again the camera introduces the performer through a series of close-up shots of their feet as they enter the translucent water. The site of resistance and complicity in the transgressive act, dancing, is the feet. Throughout *Reines*, it has been the feet that propelled bodies into action. In the last image it is again the feet that take the dancers into the lake for their final movement.

Reines as a si(gh)te-specific dance derives its powerful kinetic narrative from the projection of distanced and intimate bodies, proportioned by scale and location, onto a landscape of recognisable and real space.

The movement is specifically generated from the site but reliant on the camera and editing for its realisation.

Reines closely follows the use of filmic kineticism and transformative somatic dialogue with space/place found in the Flemish dance film and video productions of the late 1980s and 1990s.

Roseland (1990)

Roseland is a composite of three stage performances, *What the Body Does Not Remember* (1987), *Les porteuses de mauvaises nouvelles* (1989) and *The Weight of a Hand* (1990), fused into a choreographic whirlwind by Wim Wandekeybus in collaboration with fellow directors Walter Verdin and Octavio Iturbe.

Set in a dilapidated movie theatre in Brussels, *Roseland* transforms choreographic time and space by re-examining movement phrases or actions from various perspectives. Camera and editing techniques intensify, expose, elongate, compress and reposition the dancers and their movement throughout the 46-minute video.

Roseland begins with the equivalent of a family portrait. Rushing in and out of the image, frame movement is arrested – caught – then released with an explosion of energy. The credits serve as inter-titles between each sitting for the camera. The only sounds come from the screeching of the dancers' shoes and the crash of a chair used in the posing. The opening sequence after the credits brings the camera immediately into the action as it closely follows the movement of a dancer running; disengaged from time, by silencing the physical activity as the body moves out of sync with the music, and 'living colour' – through the use of black and white. Space is compressed as we view the dancer from above preventing us from experiencing any sense of volume or depth in the surrounding space while we are disjoined from the physical effort by the melodic score of Thierry De Mey and Peter Vermeersch.

Editing connects the three stage pieces together by replacing/ transposing objects that generate movement in each piece. Four means of transferring kinetic energy among the performers – throwing bricks, blowing a feather, the putting on and taking off a coat and the 'handling' of bodies – integrate body, and camera in space and time. Here dancers grab, pat, slap, catch and caress bodies and objects igniting movement exchanges that propel them and the camera throughout the space. By cutting from brick to feather to coat to hand, *Roseland* articulates the significance of the choreography, revealing 'real bodies under extreme circumstances' (Jans, 1999: 8), weaving together the bodies, objects and actions through multiplicity of intention redefining time in the process.

The cameras, always in motion, employ filmic kineticism to bring the viewer into the frame of choreographic action. In a section taken from *What the Body Does Not Remember*, the camera zooms into a hand as its body searches a female dancer. Zooming in and out we anticipate the flinching of the female body in reaction to the intrusions. An increase in, somatic awareness builds up as our proximity and culpability increases with each infringement upon her body. In reaction, the female dancers, three in all who have been subjected to body searches, aggressively stamp at their male inquisitors. As they respond with jumps the camera reacts as well with an increased abruptness and an intensification of the audio.

With the isolation of space, alternative viewing positions and frequent disjuncture caused by the camera work and editing, the viewer becomes both a participant and voyeur integrated into the piece in a way rarely possible with live performance. Vandekeybus displaces the dancer by altering the tempo, perspective and syntax of the choreographed movement. Slow motion, multiple perspectives and repetition of the same movement expand both choreographic time and space while compressing real space and isolating the screen space. The body takes full possession of the space only to be replaced by itself (repetition) or another body (insertion/transposition) affording yet another kinetic dynamic to the piece. *Roseland* in many ways fulfils Deren's motivation for 'transfiguring' the space of dance, developing 'a whole new set of relationships between the dancer and the space' (Franko, 2001: 141).

Unlike Deren's determination to create unique screened bodies and choreographies that cannot exist outside of cinematic time and space, Vandekeybus takes key frames of his choreography and choreographic process splicing them together in dialogue with the camera, music and architecture to si(gh)te the techno-presence of dance video.

Velazquez's Little Museum (1994)

La La La Human Step's *Velazquez's Little Museum*, like *Roseland*, is based on a previous stage production. It is not a re-staging of dance work, but a surrealist fantasy injected with choreographic vignettes from the 1991 piece *Infante C'est Destroy*.

Bernar Hébert imagines on film a private museum dedicated to the paintings of the Spanish baroque artist Diego Velazquez. Hébert and La La La Human Steps bring the seventeenth-century painting to life in a surrealist dreamscape infused with the passionate bodies of Louise Lecavalier, Sarah Lawrey and Sarah Williams. Hébert focuses on two paintings of Velazquez's for the majority of the fifty-minute film: *The Maids* and *Venus at Her Mirror* as the film plays on mirrors, reflections, transparencies and water images.

Velazquez opens with isolating scenes, as did *Reines* and *Roseland*, of a woman wandering down a country road at night. For the first five minutes the camera follows this woman in red, Markita Boies, as she makes her distraught way to a museum in the woods. She encounters various obstacles on her way of symbolic significance. Markita, who does not dance or speak in the film (there is no spoken dialogue), witnesses her first dance at a clearing by the lake. The location, as with the road and woods in the previous scene, clearly set *Velazquez* in the real world. The choreography, lifted straight out of the stage production, does not make any adjustments for its new spatial environment. The dancing appears surreal, out of place and yet it prepares us for the fantastic settings we are about to enter through Markita. Once at the museum, various cinematic effects introduce the viewer to the works of art and our 'tour guides' – the dancers.

Hébert plots out a course through the labyrinth of the museum, shifting space and time through kinetic manipulations of bodies, architecture and paintings. Each serves as a portal to another realm of experience – mimicking the multiple viewpoints in Velazquez's most famous work *Las Meninas* (1656). One of the most dramatic transformations of space follows an entangled duet when the room itself turns counter clockwise. Louise Lecavalier's partner, Rick Tjia, drops her into a framed opening on the floor. She falls through the frame – an image repeated in various forms throughout the film, into a tank of water. The camera captures her initial splash then cuts to her now nude body swimming into the frame of an underwater scene of *Venus at Her Mirror* (1644–48). Venus follows Lecavalier back through the frame where they perform a duet in the nude.

Figure 4   Velazquez's Little Museum Film Still (Courtesy of Ciné Qua Non, Inc. Director Bernar Hébert)

Entrances and exists from one place to the next, from one choreography to the next, are frequently facilitated by a cinematic transformation of the space. The body remains present, with the space changing. Hébert retains the postmodern nonchalance of Lock's choreography as he imbeds the choreography into the paintings of Velazquez.

Lock's choreography since the beginning of his dance company in the early 1980s has been driven by the superhuman dynamics of Louise Lecavalier. *Velazquez* is no exception. Lecavalier's dancing redefined physical theatre in the 1980s. Her high velocity spiralling that turns incessantly, throwing herself against her partner, earned her admiration and awe from many. She forcefully engages her dance partners, the performance space, the camera and the audience, complicating her reception and agency by unabashedly exposing her sexual as well as kinetic body. Her extreme physique has prompted considerable discussion concerning her gender identification. Ann Cooper Albright argues in *Choreographing Difference* that Lecavalier's 'vaulting back and forth across the stage creates an intense physicality that both literally and figuratively *crosses over* [emphasis hers] gender norms' (Albright, 1997: 29). In filming Lock's *Velazquez*, Hébert faced the challenge of capturing Lecavalier's and her fellow dancers' kinetic volume and speed. To do this, Hébert, as Verdin before him, elongates the visual line of the dance zooming in to a point where the visual has the potential to turn tactile. At first the camera keeps its distance (I speak as if only one camera is

present, but obviously we have many angles with many possible relationships); then it joins the action increasing the audience's kinetic,
even choreographic, experience. We do not hear the sound of the camera
or its operator nor do we see her/his sweat. As mentioned before, the
camera in this work does not become an extra performer – instead
I posit that the camera collaborates with(in) the kinetic vocabulary of the
choreographer and dancers bringing the viewer into the dancers kinaesthetic space. The editing must match the rapid tempo of the movement
to recreate the experience of Lecavalier's dancing.

Does the transformation of space by the camera and editing table in
dance film enable a kinaesthesia that would be unrealisable on stage?
Or is it the viewers' understanding of, or somatic reaction to, the choreographic material that becomes altered/transformed via the cinematic
intrusion into the mise-en-scène? Cinema/video allows for the transformation of time and the transfiguration of the moving body in space.
It transports bodies to a multiplicity of environments not possible with
live performance.

For Deren the editing table is the final choreographic step in dance
film. No longer is the concern with generating or capturing movement,
but with locating the movement in space and time. The presence of
the performer remains constant due to the choreography, even though
the location is in flux. In *Reines* as the location shifts it is reflected in
an adjustment of the choreography, in *Roseland* the shifting presence
is established through repetition and replacement of bodies and the
four choreographic elements (feather, brick, coat and hand). *Velazquez*
moves both dancer and space (turning of the room or exchanging air
for water), redefining the gravity of the space, moving the dance into
other planes.

Temporal shifts further construct the techno-present body by freeing
it from real time. Compression of time takes place on two levels – the
actual piece is shorter than the stage version (*Roseland* and *Velazquez*)
and the real time of watching events unfold can be compressed through
editing (*Reines* and *Velazquez*). Slow motion and repetition have the
reverse effect. We have seen examples of time expanded in the opening
sequence of *Reines* and *Roseland*.

The continued presence of the moving body in dance film – a clearly
corporeal body actively engaged with the location and locomotion
produced by the film/video incite/insights kinetic dialogue between
stage body/screen body and the spectator/viewer. Not all dances put on
film or films made as dances create such an environment for discourse.

But in the narrow genre I have defined as dance film contingent on their use of filmic kineticism I believe necessitates such an environment.

It is my strong contention that the body remains pivotal to the dance film, in part due to the complex mise-en-scène of physical theatre where the embodied architectonics relies on the material body for expository meaning.

Notes

1. *One Day Pina Asked Me 1983* made for TV, INA, Antenne 2, R.M. Arts, RTBF, SSR. 57 minutes.
2. The following argument is derived from my article 'Filmic Kineticism in the Work of Dv8 Physical Theatre', *Carbone 14* and La La La Human Steps. *Body, Space & Technology*, Brunel University, 2002.

References

Albright, Ann Cooper. (1997). *Choreographing Difference: The Body and Identity in Contemporary Dance*. Hanover, NH: Published by the University Press of New England [for] Wesleyan University Press.

Doane, Mary Ann. (2002). *The Emergence of Cinematic Time: Modernity, Contingency, the Archive*. Cambridge, MA: Harvard University Press.

Dodds, Sherril. (2001). *Dance on Screen: Genres and Media from Hollywood to Experimental Art*. Houndmills, Basingstoke, Hampshire; New York: Palgrave.

Franko, Mark. (1995). *Dancing Modernism/Performing Politics*. Bloomington: Indiana University Press.

Franko, Mark. (2001). *"Aesthetic Agencies in Flux"*. Maya Deren and the American Avant-Garde. Edited by Bill Nichols. Berkeley: University of California Press.

Jans, Erwin. (1999). *Wim Vandekeybus*. Kirsch Theater Lexicon. Brussels: Vlaams Theater Institute.

Taussig, Michael T. (1993). *Mimesis and Alterity: A Particular History of the Senses*. New York: Routledge.

Filmography

Le Dortoir/The Dormitory. (1989). Dir. François Girard. Chor. Gilles Maheu and Danielle Tardi. Rhombus Media.

Le Petit Museé de Velazquez/Velazquez's Little Museum. (1994). Dir. and book by Bernar Hebert. Chor. Édouard Lock. Ciné Qua Non.

Reines d'un Jour. (1996). Dir. Pascal Magnin. Television Suisee Rommaine.

Roseland. (1990). Dirs. Walter Verdin, Octavio Iturbe and Wim Vandekeybus. Chor. Wim Vandekeybus, RM Associates.

4
Saira Virous: Game Choreography in Multiplayer Online Performance Spaces

Johannes Birringer

Level Up

Dance and games may not appear to have a natural connection in Western culture, at least not in our historical understanding of concert dance and choreographic practice. But in discussing contemporary dance and its relationship to interactive technologies, it is tempting to look at video or computer games, their global reach in the market of a burgeoning games industry with players worldwide numbering in the millions, and particularly those games (and their social networks) being played together online. It is common knowledge that multiple-player online games, sometimes referred to as massive multiplayer online role-playing games (MMORPGs), are community-oriented and generate active player communities or clans that attract attention from researchers concerned with social interaction and group dynamics in artificial environments.

The connection I want to sketch here concerns dance which has moved into the telematic, networked terrain and shares the digital design technologies that underlie the creation of artificial environments, even though telematic dance does not aspire to be a computer game. The participation in a game world can be understood as a form of navigation in a 3D virtual environment, yet the player, of course, maintains a corporeal presence and 'plays' throughout the game with her or his body, thus involving cognitive and sensorimotor processes that are active in any engagement of spaces that can be heard, felt and intuited with our bodily intelligence. A more provocative area of research opens up if we now imagine the proprioceptive, cognitive and physical processing of temporal and spatial experience, as it is fundamental for dance, move to the foreground of performance involving participatory game scenarios or dramaturgies constructed through interactive digital technologies.

In the following, I will first describe interactive constellations which emerge in contemporary dance media and have been applied in networked performance and the programming of responsive environments. These environments will then be examined through the lens of interactive online computer games and the collaborative culture of multiplayer experience. In my case study, the game-like environment acts as a 'dance studio' where various levels are engaged by players who interact with each other and the streams. The notion of 'choreography', therefore, undergoes a conceptual reorientation as it is applied to live webstreams generated as movement with digital processes, images and sound, graphics and chat. Such choreography, in accordance with the language of computing we use in programming real-time interaction, will have to be generative and emergent rather than fixed. The physical interaction depends on input and player feedback, which in turn continuously modulate the emergent choreography. The project's conceptual development owes much to recent cross-currents in game design, interactive media art practices and telematics, while its focus on physical performance inside programmable environments expands compositional ideas in the field of dance and technology. But it also offers suggestions to game designers and to artists who recontextualise game culture by using sampling, machinima, mods and console-based subversions, as our notion of generative choreography explores performance in virtual environments which are dynamic, non-deterministic, and variable in terms of navigable space, zones, levels and role-playing, and therefore yield a different approach to game constraints created by rules and goals.[1] The rules in *Saira Virous* are entirely fictive or poetic: they offer no real constraints.

Interactive environments

There are various ways to trace the history of interactive art and the more recent evolution of interactive computer technologies which have become attractive to dance and performance artists. Aesthetic theory has generally foregrounded the parallel lines of 'interactivity' in the visual arts and sound art (installations, participatory artworks, digital sound and video sculptures, computer music, etc.), while Internet-based art, electronic writing, mobile and locative media practices, games and virtual world design only now gain recognition in various interdisciplinary research fields. Choreographers, having joined forces with musicians, programmers, and digital artists over the last decades were noted for their readiness to explore computational processes, data representation and virtuality, as movement capture and abstract motion, unlike the text-based dramaturgies of the theatre, seemed more easily compatible with data manipulation within contemporary models of emergence. Such models, influenced by the discourses of artificial life, cybernetics, AI, computer science and biology have largely replaced dramatic or realist models of representation. They undoubtedly offer a stronger stimulation for interdisciplinary artists.

Beyond the basic human–computer interaction (HCI) common to all computer systems and consoles, the design of interactive technologies today involves pervasive dimensions (networks, mobiles, wearables). In dance, interactive design obviously includes the body with all its transactional capabilities. Bodies provide data and wear sensors, and widespread attention now is given to the precise role of 'gestures' that control human–computer interaction. Performance interactivity refers to all programmed environments in which the interaction is emergent, dynamic; the interactive relationship generally involves the control of – and reaction to – digital image and sound generation or animation, the mutation of media forms. Performance, in this sense, leaves the theatre. It joins the creative industries, the Internet and streaming media, and thus is no longer site-specific. It uses techniques of moving image production or data manipulation on a microcellular level. On the other hand, since so much interactivity involves media output (video, audio, animation), performance appears glued into screen-projection media. Practically all photos we see today of such dance performances show the dancer in front of, or surrounded by, digital projections.

In telematic performances with distributed action, where images and sounds are created not simply to be transmitted from one location to another, but to spark a multidirectional feedback loop with participants in remote locations, the site of the body is a transactional collectivity: fluid, transitory, ungrounded. What emerges here is a new composite form of human-machinic performance as the streaming media is produced by physical action. In previous writings I proposed a categorisation that might help to distinguish various types of interactive environments: 'sensory environments' (based on sensors or motion tracking and an evolving dialectic between artificial world and human agents); 'immersive environments' (Virtual Reality-based, e.g. the 'Cave' or panoramic installations that integrate the body, via stereoscopic devices, into the polysensual illusion of moving through space); 'networked environments' (telepresence and telerobotics, allowing users to experience a dispersed body and to interact with traces of other remote bodies, avatars and prostheses); and 'derived environments' (motion-capture-based re-animations of bodily movement or liquid architecture, which can also be networked and reintroduced into live telepresence or telerobotic communications between remote sites). The parameters of these environment types can be mixed; we then speak of 'mixed reality environments'.[2]

In the present context, it is interesting to note that for many dance technologists the relationship of body to remote body, virtual body, avatar, or various prosthetic or data extensions of physically generated synergies was not troubled by humanist and realist anxieties at all. On the contrary, early fascination with the LifeForms software, which allowed a completely denaturalised re-composition of a movement skeleton, followed by the initial impact of motion-capture-derived animation with its immense potential for graphic abstraction, soon encouraged choreographers and animators to think more about motion analysis and motion graphics within a 'posthuman dynamic system space', and less about character, role or expressive faculties of a mimetic body.[3]

This interest in abstraction, denaturalisation and media-computational investigation, which we witness in many performances that integrate 'virtual bodies' in the choreography (e.g. Merce Cunningham, Bill T. Jones, Wayne McGregor, Company in Space, Troika Ranch, Trisha Brown, Yacov Sharir, Philippe Decouflé, Isabelle Choinière, Carol Brown, Christian Ziegler, Nik Haffner, Sarah Rubidge, Thecla Schiphorst, and many others), tends to contradict the general emphasis on 'realism' in video and computer games where the tradition of action games has favoured, especially in the popular 'first-person shooter game', a clear relationship between the player's viewpoint (performed by an avatar) and the 'figures' in the virtual game world.

There is an emphasis in the mainstream game industry on designing interfaces which construct a direct view of the screen scene through the player's own eyes, matching and synchronising player actions with changes and movements on the screen, thus producing a sensation of being 'in the scene', regardless of whether it is a hardcore shooter game like *Halo* or role-playing game/quest games like *Sim City* or *Ever-Quest*. It has often been noted that the environments of these games are rendered in highly realistic detail, with naturalistic surface texture, dramatic lighting and sound effects, and subtle use of colour. As Andrew Darley and others point out, the 'majority of adversaries – monsters, zombies, aliens and so forth – are rendered and animated with the same high levels of surface accuracy and increasingly this is combined with a persuasive anthropomorphism'.[4]

The myth of interactivity

Before one gets carried away by the glamour of the (much abused) notion of interactivity, it may be necessary to distinguish complex interaction from a shallow clicking of buttons or triggering a reaction from a set sequence programmed for users. The dexterity of a game player using a console, like the technical skill of a dancer performing with sensors, does not tell us much about the aesthetics or immersive experience of the game or the dance, nor about the qualitative input the player/dancer may have in the creation of visual, auditory, synaesthetic and narrative events, much as surfing or browsing the web, clicking through a series of screens and links or engaging a database indicates little more than a functional interface common to interactive product design.

As with all commercial interaction design based on the conceptual paradigm of the desktop (CPU, monitor, keyboard, mouse, pad), and even with the new generation of ubiquitous, pervasive or wearable computing which promises tangible and embedded interfaces integrating computational augmentation into the physical environment, the cultural and artistic rhetoric surrounding interactivity has been high-minded and often misleading. The world has not become a better, more democratic place, participatory design is rare, and interactive art has not necessarily made the 'user' a co-author nor allowed the user-player the kind of active role and freedom of expression that is implied in an interactive exchange involving autonomous development.

We have also seen a good deal of interactive dance which displayed the impoverished cause-and-effect triggering of responses activated via motion tracking, infrared, pressure or heat sensors and accelerometers. The embarrassment of interactivity is particularly pronounced in interactive installations that involve more complex, sophisticated programming but address an unaware, untrained audience-user who is reduced to gesticulations and tip-toeing, or a prolonged period of trial and error, trying to figure out how the work works or what it allows one to do. It is instructive to walk through the ZKM Media Museum in Karlsruhe and see the evolution of interactive media art, comparing the clunky, the intuitive, the sensual, the coldly mechanical, and the enervating interfaces that leave your imagination depleted. How many interactive installations have you entered that engaged your fantasy and creativity? The myth of participation, thus, has to be measured against the notion of 'pleasure' that is used by games analysts who argue that genuine interactivity needs to meet some conditions for successful play and an increasing attachment to a game.

The pleasure of gaming and telematic pleasure

In terms of a general definition of what constitutes a video/computer game, and without entering a larger cultural, philosophical and sociological study of game culture and *homo ludens* (Huizinga), one might propose that 'game' or 'play' creates a separate frame from the real world. Rules and behaviours apply to this frame. Most computer games have a theme, that is a subject matter that is used in contextualising the rules and the player procedures and mechanics they allow. The game theme provides a meaningful context for everything that takes place in the game world. Game environments provide the space for components, procedures, narratives, and actions. It is obvious that the game environment is critical, as it is specific to each individual game. In its defining features, a computer game can be described as 'a rule-based formal system with a variable and quantifiable outcome, where different outcomes are assigned different values, the player exerts effort in order to influence the outcome, the player feels attached to the outcome, and the consequences of the activity are optional and negotiable'.[5]

While the attraction of games may reside in their theme, the characters, and the aesthetics of the game environment, it has been widely argued that the immersive quality of gaming resides in the player activity. A computer game has to challenge, it must provide exciting situations to experience, stimulating puzzles to engage with, and interesting environments to explore. 'Gratification', James Newman suggests, 'is not simply or effortlessly meted out'. The pleasure of such play, furthermore, 'is derived from the refinement of performance through replay and practice. Consequently, it is essential that obstacles, irrespective of the form they take, must be "real" in that they must require non-trivial effort to conquer them.'[6] The centrality of participation and the sense of 'being there' also suggests that players demand interaction in order to effectively feel they are enacting or role-playing the fantasy, moving towards gain or loss. Newman notes that 'it is the primacy afforded to doing and performing that renders "non-interactive cut-scenes" so unappealing to players', as such movie sequences prevent direct control or interactivity.[7]

The question of 'doing' could be elaborated if we looked at different types of environments and spatial narratives, since the cognitive processing of the game environment and its challenges, for example in its investigatory or exploratory dimensions, especially so in the collaborative multiplayer games, will vary from case to case. Communications between players in multiplayer games, for example using voice-over IP tools rather than the older text-based forms, build a more vivid sense of mutual awareness, attention and enjoyment. What is particularly relevant for our contextual shift to collaborative telematic performance is the fact that the pleasure of gaming excites the body, and that much current research shows how multiplayer online games also generate 'affective alliances', social formations, on and offline information networks, and communal relationships for the sharing of resources.[8]

The pleasure of telematic dancing is based on very similar processes of socialisation and self-organisation, and along with my fellow members of The Association of Dance and Performance Telematics (ADaPT), I believe that after five years of shared online performance our group has formed a digital community.[9] ADaPT's practitioners come from diverse backgrounds but share an interest in performance, new media, and collaborative experimentation. Together, the artists and technicians have developed hybrid models of networked interaction that combine learning through the iterative pattern-recognition of dense multi-layered environments (physical and virtual). Most importantly, acting in such mixed-reality environments has not just challenged our body-perceptions and proprioceptive awareness, it also requires a very specific studio architecture and knowledge of tools allowing for the interplay of dancer, 'physical camera' (a camera that can dance), precise lighting, sound distribution, and the computing environment which includes various interactive or global media controllers. Dancing across different times zones in international collaborations which also involve writing and chat-windows not only raises practical questions, namely determining degrees of collaborative agency in distributed, multi-user platforms, exploring perceptual learning curves and organisational strategies. It also stimulates our pleasure to experiment with playful scenarios that would not have occurred to us on a theatre stage.

The plasticity of the telematic game

The telematic computing environment involves both hand-held physical cameras as well as a video/computer-controlled movement sensing system. The global media controller organises the sonic and graphic output for the sensing system. It is an instrument that primarily controls the source materials (sound and video files stored in the computer or synthesiser), sound parameters, and the dynamics of real-time synthesis which occurs between dancers/performers and physical camera. The environment can harbour considerable complexity since the software patches (for example, in MAX/MSP/Jitter, Isadora, softVNS, Eyecon, BigEye, etc.) can be constructed in the manner of a 'nested' design – enfolded entities that are in a continuously fluctuating state of unfolding to activate the modular parts. Moreover, interactive multi-media performance generally uses interconnected systems (linking the computers through the network) to drive several patch programmes for sound, video, and motion tracking simultaneously. The studio archi-tecture needs broadband network access points so that outgoing and incoming webcasts can flow. The streams can be imagined as digital video projections; in our studio in Nottingham we use multiple screens (suspended from ceiling to form a curved or angular space) for large, human-size projection. All instruments and tools for the real-time gener-ation of the streams are inside the space, close to the performers and operators. We also use surround sound, and in a telematic performance there usually are multiple channels of streaming audio in operation.

Given such complexity in the programmed and scenographic envir-onment, we might ask how performers, video artists, musicians, and programmers regard the physical relations and plasticity between performance and controllable parameters, and how dancers can see their movement as a form of topological mapping of the body's experience and proprioception within the interface. This question, at the same time, allows us to speculate on the particular spatiality of the intermediated player/world we are constructing for the performer who enters 'into' it. In terms of game theory, this would be the point of entry that creates the affective bond between player and game world. In telematic terms we can call it 'immersive connectivity'. The programming goal is to integrate an image-based recognition system (e.g. a computer running MAX or Isadora) or a motion sensor interface (e.g. the MidiDancer or I-Cube sensors) into a unified MAX, VNS or Isadora environment.[10] The 'technical' integration implies that the dancers understand the system underlying the theme of the game and integrate parallel parameters into their movement intelligence, their increasing awareness of tactile image projection spaces (as we use them in extreme close-up scen-arios for telematic performance). How do dancers dance with images, avatars and monsters? From a choreographic perspective, the dancer within an interactive environment familiarises herself with the response behaviour of the sound and video parameters, and with the immediate stream-image that the physical camera generates of her close-up. Here the dancer is taking in at least two image fields, one of her own locally generated stream, the other of the incoming stream(s).

Dancers and programmers will strive to create an exponentially more sensitive, articulate and intuitive system, and the role of the physical camera operator is very intimate, to the extent that I see dancer and camera as a duo. At the same time, the dancer creates data: in a shared environment this also requires of the programming a constant, iterative designing process, refinements in sensors, filters, and output processors, while it suggests an attenuation of the performer's spatial-temporal consciousness. If she wears multiple sensors, as well as acts in a duet with the camera, she needs to do a great deal of parallel sensorimotor/cognitive processing, developing an extreme awareness of her movements and which movements control what parameters. Her gestures tap into tactile, spatial, acoustic and kinetic sensations that go beyond the immediate kinesphere of the body. If she performs in conjunction with video or 3D projections, she animates these virtual image-spaces that are constantly emerging and also unpredictable. Her movement, if it affects and responds to the image movement, crosses real and virtual spaces. She moves in-between.

Noticing how open, complex, and evolving this networked scenario is, we became interested in the function of playful rules or constraints as a way to invent an interactive dramaturgy based on a 'game engine' rather than prolonged improvisation which proved ineffective in online collaboration. At that time (2003–04) I had joined a group of gamers in Nottingham who organised the annual ScreenPlay festival featuring independent games and interactive art. Many of our discussions at the festival focused on the relationship between games, play, ritual, fun and learning – question raised in equal measure by game developers, educators, psychologists, and artists using reverse engineering to subvert game consoles. I had also become aware of Blast Theory's efforts to take games into the street to test the notion of ubiquitous computing in streetplay.[11]

When I developed the dramaturgy for *Saira Virous*, jointly with our ADaPT partners in Arizona, I used 'game engine' as a metaphor for a scenario in which players enter online to work with remote partners in an exploratory journey. I wanted to make the entry a physical journey based on fictive tasks, whilst our partners in Arizona, who named their part of the game *Viroid Flophouse*, chose to work with an iconic 'board game' designed to look like a grid labyrinth mapped with hieroglyphic signs. In the concluding pages, I describe the different dramaturgies of the game and try to theorise the notion of a telematic dance game.

The plasticity in the telematic architecture distinctly intertwines numerous agencies (gestures, bodies, machines, media flows, data): the dancer moves in real space but also enters into screen space, her corporeal action is digitised, her voice is transmitted and received (it returns as echo, with a slight delay) along with the responding stream and the fluid images generated by the partner site(s). The interactive environments allow the real-time synthesis of various media forms (video, audio, text, graphics, data transfers via MIDI affecting directly the processing/configurations between sites), but this synthesis is part of a larger synaesthesia. The technical 'system' is only a feedback system in which information travels and transduces bodily activity into computation which controls the instant media outputs (sound, video). But this system is not an environment that affects proprioception, vision, hearing, smell, and haptic sense unless we understand the sonic and the projected images, in their behaviours, as transformative, acting upon the sensing body and intensifying unaccustomed connections, orientations, deformations, or interferences with the functioning of the body that now, and continually, must adapt to the unpredictable conditions, the dissipative states.

For example, I wanted to include a scene in the game in which the players perform with eyes closed, activating bodily modalities apart from sight. The process (proprioception, tactility, affectivity) through which human perception constructs 'images' does not depend on the visual mode, and telematic dance is not directed primarily at an audience watching the scene, but at players enacting a fantasy.[12] Entering virtual worlds opens up a perspectival flexibility, and our response to the surface, the grain of sound and the skin of the images intensifies. Thus we could almost argue that we 'wear' the filmic-projected textures differently, that we touch and sense the video images with our whole body.[13] This 'wearing the digital' challenges choreographers to think of the relations between dance and projected image in different ways.

For *Saira Virous* I sampled and remixed a few moments from the language/plot of David Cronenberg's *eXistenZ* (1999), a film which opens with a designer's live demonstration of a new virtual reality game, except that unlike the video games of our present day reality, the games in *eXistenZ* are delivered directly into player's nervous system via the metaflesh game pod, a surreal coupling of amphibian nervous systems and technology. Things go wrong during the test, and the film's plot, similar to those of *The Matrix* and *The Thirteenth Floor*, develops its ever more perplexing confusions of reality-within-reality-within-simulation. Its most interesting aspect, the melding of technology and biology, had an almost viral effect on our imagination, especially as we were intrigued by very quick 'cuts' (level changes) between scenes. The biomorphic contorsions or grotesque torque in which the players find themselves resembled the digital motion animations we had worked with in our interactive rehearsals. We proceeded to invent our own dramaturgical devices and set up a level structure which potentially can be extended much further:

1. Welcome by MC.
2. Second intro by Mystery Figure (Saira) preparing Players for training session.
3. Level One: Moving in all directions: boule game/rolling dance.
4. Game-journey begins.
5. Level Two: Obstacle – game only works when you double up.
6. Level Three: Intimacy Blind scene, listening to Inner Voice.
7. Level Four: Final Fantasy arrives for kosupure (dress change), character must change identity.
8. Level Five: 'Infection' starts mutation process. Texas desert landscape. Players must swim across the channel to reach ocean.
9. Saira must tell players whether she loves them or not, otherwise they lose/drown/dry up and shrivel. Love and infection go hand in hand. Hand dance.
10. Level Six: Obstacle by Particle Systemics (Lycanthropic movement: hands dematerialise).
11. Level Seven: Playing with very small object (kawaii). Location changes to underwater.
12. More Bad Things surely happen.

The subsequent rehearsals and written exchanges between partner sites (Tempe, Amsterdam, Detroit) developed the levels and changed directions. Our Arizona partners introduced the idea of self-sacrifice: the gamers must express their devotion to the obstacle (virus) through the act of self-destruction. The environment is a metaphor for the gamer: The gamer must destroy or dissolve the environment within specified parameters (motion sensing triggers). The virus then can accept or reject the act of self-sacrifice based upon these parameters. In the event of acceptance, the gamer is regenerated and proceeds to the next level. In

Figure 5   Saira Virous, 'Blind/Inner Voice Scene', telematic dance game, ADaPT, 2004 (Framegrab: Johannes Birringer)

the event of rejection, the gamer goes back to the first level. *Flophouse* is a pun on the idea that in gaming environments death of a character does not constitute non-existence. The character is regenerated by starting the game over again.

In *Saira Virous*, Nottingham and Detroit used an 'empty space' in which all visual and sonic elements were created on the fly using live camera work and sampling (prerecorded media) to generate the immersive context for the journey from level to level. The navigable space is constituted by the image-movements the partner sites enact through affective bodily interfaces. The Arizona and Amsterdam sites both had the same stage configuration. The first level of the game consisted of an image of a sculpture garden. Both gamers had to step 'through' the door, thereby triggering the motion capture region associated with that area, in order to move to the second level. When the movers in one space occupied a square within the region map, that area was highlighted on the game board in the remote partner's space.

Each square had a distinct and characteristic sound that accompanied the movement of the game piece (i.e. the performers). This sound was shared between the sites. When a hidden square was activated, a special sound indicated to the performers that they had found one of five hidden squares. Once all five hidden squares were uncovered, the board changed to one of five winning messages and the game was over for those participants.

Figure 6 Viroid Flophouse, showing gamers inside 'hieroglyphic region map'. ADaPT, 2004 (Videostill: John Mitchell)

The original idea for the viral fiction was for remote sites to infect each other using digital telepresence. Initially the performance took the form of a cooperative game, where players on each side of the Atlantic either learnt a specific fictional task that allowed them to invent a role or used motion sensing to control a graphic environment in the partner space. Over time the game engine was developed, based on the action of the remote and local gamer 'sharing' the same space. In *Saira Virous*, the gamer in Nottingham and the gamer in Detroit learnt to navigate surreal landscapes with unconventional movement or perceptional tasks, such as moving only with one arm, listening to a song inside the head or swimming across the channel to Japan wearing goggles. In *Flophouse*, the gamer in Arizona and the gamer in Amsterdam had to position themselves in the same virtual gaming space in order to get to the next level. The performers then worked together by means of their virtual location in the performance space to cooperatively solve the puzzle in real-time. A master of ceremonies in a third location (Nottingham) made casual observations, gave encouragements and subtle warnings to the performers throughout the performance.

What I have described is the beginning of a multi-phase project shifting between 'collaborative' and 'competitive' forms but primarily exploring the physical and programming dimensions of such online multiplayer environments. The game is a useful step for us into the realm of augmented reality using multi-modal performer interaction and information streams to connect sites over distance in large-scale, movement-based works. One important aspect of the work is that it is participatory: it really only makes sense to be involved in the game, rather than sitting down to watch it as a spectacle. In this sense, the notion of choreography applies to the interaction design and the ideas for the game engine, allowing not only trained performers but audiences to engage in the playworld. It is a physical playworld which cannot be experienced at a console or computer but actually needs to be entered literally.

This is precisely how I understand telepresence – to be present in a distant image world which is being created as I become present in it. Here we also observe the particular challenge of camera work (framing, angle, motion) within and against the frame compositions of the virtual space. *Flophouse* already had its virtual map designed, while it does not exist as a world until the physical game gets under way and the role-playing movers and their actions are transmitted via cameras. Digital real-time dramaturgy implies that the dancer is integrated into – or inserts herself into – a moving architecture generated with real-time data and samples.

Telepresence or tele-action also means entering into the streaming images of the remote site, thus affecting the reality or virtuality perceived at that location. It is crucial to recognise that live performance, unlike synthetic computer-generated environments, brings corporeality as real material into the teletechnologies and, via the streams, to a real remote physical location. Another critical difference, for example between telematic dance and telerobotics, is the motivated physical action which travels from one location to another: it operates on the remote action and changes its reality. Since it is not artificial, it can understand irony. This perhaps distinguishes telematic dance from MMORPGs and the priority of 'rewards' in commercial games. I can play with the distanced body images and have ironic relationships with the processing of my movement and the fictive rules it breaks, enjoy the thrill of the exchange of energies and strange fantasies with performers in the other sites, and savour the natural precariousness of temporary networks with their lags, interruptions, and collapses. These network environments, after all, behave like the weather. The materiality of its temperatures makes such game technologies exciting: to be streaming together teaches us a great deal about living systems and the humorous aspects of the 'control' of a game. In telematic dance, we like to lose control.

Acknowledgements

I acknowledge the contributions of the members of the Nottingham LATela team (Lori Amor, Marie Denis, Wayne Green, Takuya Inaba,

Richard McConray, Mareen Ménigault, Rosie Laws, Shih-yun-Lu, Helenna Ren, Meg Rowell, Rohane Renton, Yoriko Saito, Laura Wolfisz, Ben Walker and Peter Bowcott) and thank the collaborative culture of the ADaPT team members, especially the Arizona State University Department of Dance.

Notes

1. Sampling and modification, hacking into game consoles and appropriations of game engines to create machine cinema ('machinima') are emergent hybrid forms through which the aesthetics and the technology of video games are adopted and subverted, as we have seen in music and hip hop culture as well over the past decades. Thus we also see the close links between game cultures, club cultures, Dj-ing and Vj-ing. In a broad sense, Paul D. Miller (DJ Spooky) sees such derivations and re-mixes as part of the general 'rhythms' of sampled culture (Miller, 2004). In DJ Spooky's own mix tape, *Riddim Warfare* (1998), he samples and manipulates sounds from Atari games and fuses them with a range of other cultural references. For an excellent introduction to current games research, see Marinka Copier/Joost Raessens' anthology (2003) published after the inaugural conference of the Digital Games Research Association (DiGRA). The second international conference, titled 'Changing Views: Worlds in Play', took place in Vancouver (2005) and was attended by many hundreds of artists, designers and researchers attesting to the growing presence of games research linking industries to the academic and educational sectors.
2. Birringer, 2004a, 96; Birringer, 2004b and 2006. The mixing of parameters occurs ever more frequently, as all interactive softwares can potentially be linked. As an example of research into pervasive computing, Steve Benford's team at the Mixed Reality Lab (University of Nottingham) became known for its collaboration with Blast Theory, a performance group exploring city-wide gaming projects (*Uncle Roy All Around, Can You See Me Now?*). Other recent experiments include Active Ingredient's *Ere Be Dragons*, a game where players go on a journey with hand-held devices, and as they explore the landscape, another virtual world is created by their own heartbeat.
3. Sampling N. Katherine Hayles' theories of the 'posthuman', Marlon Barrios Solano offers a provocative thesis on the relations of improvisation to inter-active design (Barrios, 2005).
4. Cf. Darley, 2000: 30. But see Newman, 2004: 121ff, especially his comments on non-contiguity in videogame space.
5. Juul, 2003: 35.
6. Newman, 2004: 16–17.
7. Ibid.: 17.
8. Cf. Lin, Sun, and Tinn, 2003: 288–99.
9. The ADaPT was founded in 1999 linking sites in Tempe (Arizona), Columbus (Ohio), Salt Lake City (Utah), Madison (Wisconsin), Irvine (California), then expanding to include Brasilia and São Paulo (Brazil), Detroit (Michigan), Nottingham (UK), Amsterdam (The Netherlands) and Tokyo (Japan). Website: http://dance.asu.edu/adapt/ (accessed 12 December 2005).

10. Mark Coniglio, a musician/software programmer known internationally for
     his work with Troika Ranch, a New York City-based company he
     directs with choreographer Dawn Stoppiello, wrote two interactive
     programmes, Interactor and Isadora, which map data input to control
     a variety of media outputs, for example sonic, video, lighting, and
     robotic (http://www.troikatronix.com/). The other main interactive software
     mentioned is MAX/MSP/Jitter. Jitter is plugged into Max/MSP and comprises
     a set of 135 video, matrix and 3D graphics objects for the Max graphical
     programming environment, extending the functionality of Max/MSP in a
     flexible way to generate and manipulate matrix data – any data that can
     be expressed in rows and columns, such as video and still images, 3D
     geometry, but also text, spreadsheet data, particle systems, voxels or audio.
     A software like Jitter, similar to Supercollider and other Vj-ing software,
     is vital to anyone working with real-time video processing, custom effects,
     2D/3D graphics, audio/visual interaction, data visualisation and analysis.
11. The challenge for Blast Theory existed in the mixing of pre-programmed
     game content with live performance and behind-the-scene orchestration.
     When I witnessed *Can You See Me Now?* at the 2003 Dutch Electronic
     Arts Festival, online players were invited to play a chase against members
     of Blast Theory. These players were dropped at random locations into a
     'virtual Rotterdam'; using their arrow keys, they could then move around
     the city and also communicate with other players. On the real streets, several
     'runners' from Blast Theory – equipped with hand-held computers and satel-
     lite receivers – tracked down the online players. If a runner reached within
     five meters of an online player's location, that player was 'seen' and elim-
     inated from the game. *Can You See Me Now?* is a game that happens simul-
     taneously on the streets and online, inviting the public to investigate the
     near ubiquity of hand-held electronic devices in the general population, the
     presence of satellite and GPS systems, and the consequences of the blurring
     of discrete zones of private and public space. By sharing the same real/virtual
     space, the players online and runners on the street enter into a relationship
     both playful and adversarial. Participants rediscover themselves as *data* in
     a navigational hide and seek game. See also, http://www.blasttheory.co.uk/
     and http://www.mrl.nott.ac.uk (accessed 12 December 2005).
12. It would be interesting to examine contemporary interactive art installations
     in terms of their use of more auditory or haptic interface designs, and in
     terms of their abrogation or reliance on video and 3D visualisation. *Intimate
     Transactions*, an interactive installation created in 2003–04 by the Australian
     Transmute Collective (Keith Armstrong, Lisa O'Neill, Guy Webster *et al.*) and
     recently shown at the New Territories festival in Glasgow, is a case in point
     where the interface design is unusually thoughtful and challenging from a
     synaesthetic perspective. Two participants situated in different physical or
     geographical locations will simultaneously interact with the work, reclining
     within in a new form of furniture (bodyshelf) that detects their bodily move-
     ments (feet and back of body). Each participant generates flowing combin-
     ations of digital imagery and sound, including ghostly, ethereal performing
     bodies (avatars), dynamic sculptural texts and immersive sound textures,
     by moving their feet on a floorboard or rubbing their back and shoulders
     against the shelf. Using the physical interface, gently moving their bodies

on the 'smart' or responsive surfaces, participants work both individually and collectively to generate a 'world'.

13. My reference to 'wearing the digital' is based on a new project I have developed with fashion designers in Nottingham, where we explore the relations of fabric design, soft technologies and interactive design/performance. One aspect of this work includes exploring wearable computing and the inclusion of biofeedback sensors in the weaving of the textile design which can allow a direct influence on the projected environments around the body through tactile self-reference. The notion of 'wearing' film is drawn from Gaines, 2000. See also Marks, 2002 and Massumi, 2002.

References

Barrios Solano, Marlon. (2005). 'Designing Unstable Landscapes: Improvisational Dance within Cognitive Systems', in Johannes Birringer and Josephine Fenger (eds), *Tanz im Kopf/Dance and Cognition*. Hamburg: LIT Verlag, 218–30.

Birringer, Johannes. (2004a). 'Dance and Interactivity', *Dance Research Journal* 35: 2 and 36: 1, 88–111.

Birringer, Johannes. (2004b). 'La Danse et la perception interactives', *Nouvelles de Danse* 52, 99–115.

Birringer, Johannes. (2006). 'New Environments for Interactive Dance', in Nigel Stewart and Gabriella Giannachi (eds), *Performing Nature: Explorations in Ecology and the Arts*. Frankfurt: Peter Lang.

Copier, Marinka and Raessens, Joost, eds (2003). *Level Up*. Conference Proceedings. Utrecht.

Darley, Andrew. (2000). *Visual Digital Culture: Surface Play and Spectacle in New Media Genres*. London: Routledge.

Gaines, Jane M. (2000). 'On Wearing the Film', in Stella Bruzzi and Pamela Church Gibson, eds, *Fashion Cultures*. London: Routledge, 159–77.

Juul, Jesper. (2003).'The Game, the Player, the World: Looking for a Heart of Gameness', in Copier and Raessens (eds), 30–45.

Lin, Holin, Sun, Chuen-Tsai, and Tinn Hong-Hong. (2003). 'Exploring Clan Culture: Social Enclaves and Cooperation in Online Gaming', in Copier and Raessens (eds), 288–99.

Marks, Laura U. (2002). *Touch: Sensuous Theory and Multisensory Media*. Minneapolis: University of Minnesota Press.

Massumi, Brian. (2002). *Parables for the Virtual: Movement, Affect, Sensation*. Durham: Duke University Press.

Miller, Paul D. (2004). *Rhythm Science*. Cambridge: MIT Press.

Newman, James. (2004). *Videogames*. London: Routledge.

5

Artistic Considerations in the Use of Motion Tracking with Live Performers: A Practical Guide

Robert Wechsler

Introduction

Using human motion in performances to control sounds (e.g. music) and images is as old as theatre itself. Ancient Greek theatres employed elaborate mechanical stage devices to amplify the gestures of performers allowing them to portray gods with superhuman powers.[1] When King Luis XIV raised his arms in his *Sun King*, hidden strings caused the sun to rise.[2] In the 1960s and 1970s, New York City was host to 'performance art' events. Robert Rauschenberg and others began performing their art, making use of mobile, functional devices of all kinds, applied both in their intended and non-intended uses.[3] Electric eyes and other remote-acting electronic sensors were novelties at the time, but bespoke the coming high-tech digital revolution. In 1969, choreographer Bill Evans strapped portable sound boxes equipped with electric eye light sensors onto the bodies of his dancers, so that different pitches of sound were created depending on their positions in relation to the stage lighting.[4]

As digital technologies such as video surveillance cameras and computers became affordable in the mid 1990s, they were taken up by a number of technology-interested artists who saw in them a rich untapped vein of artistic potential. A handful of dance companies focused exclusively on their use, including Troika Ranch,[5] Ventura Dance Company,[6] and Palindrome.[7]

As computers, human interface devices and software became increasingly sophisticated in the 2000s, it became clear that a kind of dialogue between actor and machine is possible. The term 'interactive' became jargon. Not only do performers now relate to one another on the stage, and to the audience, but they can interact with a technological systems.

The performer's environment has been brought to life; it can seem to be responding directly to their actions.

Today, in the early twenty-first century, audiences are tiring of digital effects and the interactive performing scene is in somewhat of a crisis as it struggles to define and develop artistic applications and rationales for the use of technology in general. Artists themselves are partly to blame for this quandary as so many poorly conceived interactive works have been put on stage. Shifts in perception regarding the significance and quality of work are not only inevitable in any emerging art form, but are vital for a healthy evolution. This paper is concerned with learning from mistakes. From the standpoint of technologies used, two basic approaches are in use today: those based on physical/physiological sensors (body-worn devices, floor sensors, etc.), and those based on remote sensors (video, infrared, lasers, etc.). Both approaches give rise to unique performance qualities and both have been used effectively by numerous artists. The focus here is on motion tracking: its practical uses and artistic implications.

Definition of terms

Motion tracking

Motion Tracking, like 'Motion Capture' and 'Motion Sensing', is used variously by designers, artists, and engineers to describe systems in which video cameras are attached to computers. With the cameras focused on humans in motion, data can be collected and processed for any number of purposes. Scientists collect data in this way for research in sports, musicology, biology, psychology, and so on. Artistic applications include stage and installation art in which human motion data is used to generate or influence secondary media such as sounds, music, or projections. The expression 'motion capture' implies the recording of movement data for later processing ('capture' is a function of digital video editing systems in which a video is transferred from a digital or analogue tape to the hard drive of a computer). Systems such as Vicon (Vicon, 2006) are used routinely in the motion picture industry to create animated characters with realistic animal or human movements. The process typically involves 12–24 video cameras, set at various heights and arranged in a circle around the action. By wearing reflective balls attached to certain positions on the body, a collection of points moving in three dimensions is calculated by the computer. These moving points, which can be seen on computer screen, can be connected together to make a moving stick figure, or given a 'skin' by modelling programmes such as Maya (2006) to create animated characters for films. The Vicon company uses the terms Motion Tracking and Motion Capture somewhat interchangeably to describe its products and their features.

The term 'motion sensing' was coined by Frieder Weiß in 2002 to describe EyeCon and systems like it, arguing that this is a better descriptor since it is a system designed to give a sense of the motion, rather than exact data on position and motion. While EyeCon does indeed have a feature to allow it to monitor the positions of individual performers on stage, it is rarely used. Rather, EyeCon is an excellent tool for dancers and choreographers to control music and sound samples with their *movements* (as opposed to *positions*). It is quick to set up and, with its live video window and mouse-manipulated graphic elements, it is intuitive and easy to use. Multiple cameras can be attached, though they cannot be used simultaneously, at least not with only one computer. This may appear to be a considerable weakness in the system. It is, however, for reasons which become clear later in this article, of limited real importance.

Realtime

While the terms 'Motion Tracking', 'Sensing', and 'Capture' are clearly somewhat interchangeable, one distinction is important: some systems lend themselves more easily than others to realtime media manipulation. The choreographer Merce Cunningham made motion tracking famous when he collaborated in 1999 with installation artists Paul Kaiser and Shelley Eshkar to create *Biped*, a dance piece in which dancers appear both in the live sense, and in the form of projected animations. Since the moving figures seen on the screen are in no way influenced by the movements of the live performers, this is not a realtime system. The motion-captured data was processed in advance, and then projected simultaneously with dance.

With rare exception,[8] the highly accurate (and, by the way, expensive) motion capture systems like Vicon are not used in realtime performance settings. Not only are the many cameras and special costume requirements distracting to performers and audience, but the quantity of data is simply far more than is necessary to accomplish an interactive effect. While there is little doubt that motion capture systems will be used in the future in realtime settings with increasing ease and cost effectiveness, for the needs of most interactive performers today, systems such as Vicon are, in their current generation, unwieldy and, in a sense, overkill.

Interaction

The human body in motion draws its expressive power in no small measure from the special sensitivity of the human eye to particular qualities of movement. These are felt by the actor or dancer in her or his so-called kinaesthetic awareness, but they are equally important to the audience for these are the things which guide their appreciation of the body in motion, that is the dance. Thus, where human movement is concerned, there is a considerable difference between what the eye *sees* and what the viewer *perceives*. While we may not be aware of it, the human brain does considerable 'image processing' to the signals sent it from the retina of the eye. Sometimes the subtlest of dance movements play crucial roles in our how we feel about what we are seeing, while at the same time large movements may be ignored – essentially rendered invisible to us.

Thus it is that we must think of interaction primarily as a psychological phenomenon, rather than a technical one. One certainly does not need cameras and computers to be interactive. Interactivity is simply the instinctive back and forth of energy which occurs when animals come together to speak, gesture, touch, or, in the case of humans, create art. The heart rate and other basal metabolic indicators rise, we often tense various face muscles, and so on. Interaction is what we are *not* doing now. If you were to write me back a letter, I would have your *re*-action, but we would not really start *inter*-acting until we met up and hashed it out. Just to be completely clear on this point: automation in no way implies interaction. Bringing a new technology before an audience may make for excitement, but this not interactivity.

From the perspective of the performance setting, one may speak of different kinds of interaction:

- between artists
- between artists and audience
- between audience members
- between artist or audience member and a computer system (to the extent that the more intelligent of these systems do indeed allow a back-and-forth to occur).

But let us not split hairs. The same basic principle applies to each of these relationships. They share the same psychological roots and in practice function in similar ways. In all cases, interactivity depends on a certain degree of looseness, or openness in the artistic material, which allows for a convincing exchange to take place.

This quality of looseness/openness is similar to, though not the same as, improvisation. Palindrome's work is probably 80 per cent choreographed, that is the movements are largely fixed. And yet, even within structured material, a certain feeling of play (in the sense of clearance) is necessary in order to generate an interactive effect. If a piece is completely fixed, like television, it cannot be interactive. It would be as if attempting a conversation with someone who already knows exactly what they want to say (the exasperation of which is surely known to all).

To sum up, by this meaning of the word, interaction is a feeling you can achieve in a performance setting. It relates to spontaneity, openness and communication, and while its exact definition may be somewhat vague, we all know when it is present and when it is not. Is more interaction more artistic? Should it be the goal of the artist to be as interactive as possible? This obviously depends on the piece, but in general, of course, not. It does, however, offer a good reason to explore new ways of employing technologies such as motion tracking. Before we look at the ways in which motion tracking can affect our appreciation of a piece of theatre or dance, let us look at functionality of one motion tracking system, EyeCon (Weiß, 2005).

Figure 7   This image shows a realtime video effect linked to a technology that responds to the touch of two dancers (From the Palindrome opera *Blinde Liebe*, 2005; Dancers: Aimar Perez Gali, Helena Zwiauer)

The EyeCon system

When EyeCon was first developed, it had two elements: one allowed the presence or absence of a body or body part at distinct locations to be identified. These are the so-called 'touchline' elements. The second allows the detection of motion within defined, rectangular fields. These are the so-called 'dynamic field' elements. Beginning around 2000, it became clear that additional elements could be useful. Thus Weiß developed the so-called 'feature fields' which allow certain shape and position-oriented parameters to be measured:

- width
- height
- size
- left-most, right-most, upper-most, lower-most point the body
- degree of left-right symmetry in the body
- degree of extension, or compactness of the body's form
- direction of travel (this can be done either using the overall motion of the figure's outline or by looking at directional progress of the most-leading edge).

Input, output, and compliance

These are the input and output parameters applicable to the EyeCon system:

Input

EyeCon allows a variety of movement parameters to be used as input to the interactive system:

- Position of body parts in space around you (i.e. the touchline feature, or triggers in space).
- Movements dynamics (total body movement) within defined fields.
- Position of the body on the stage area (using overhead camera).
- Height of body from the floor.
- Width (measured from left-most to right-most point on the body).
- Degree of expansion or contraction in the body's pose.
- Size of the body image (this is not quite the same as expansion–contraction; the former is relative, the later absolute).
- Degree of symmetry in the body (how much left side resembles right side).

- Number of dancers on stage (tracking feature).
- Relative closeness of dancers to one another.
- Position tracking (location body average position on XZ grid).

Output

Audio

- musical notes (synthesisers) and samples (wave files)
  - on–off
  - volume control
  - pitch bending
  - panning.

- realtime digital signal processing
  - scrubbing forwards and backwards through sounds or text
  - complex modulations of various kinds.

Video

- on–off
- play–freeze
- film play forward–film play in reverse
- realtime video processing, for example Isadora (Coniglio, 2006) or Kalypso[9]
- realtime video DSP (digital signal processing).

Stage Lighting (including moving lights)
Mechanical Devices (mechanical hammers, wind, etc.)

Artistic implications

This section is divided into two parts: Mapping and Other Relevant Issues.

Mapping

There are three aspects to mapping: input, output, and compliance. Input and output are obvious. Compliance refers to the nature of the causal relationship: its disposition, direction, finesse, and so on. In other words, the nature of the causal relationship. Compliance concerns the psychology of the relationship: input and output. This is a complex question involving issues of visual and acoustic perception and the relationship between the two in performance settings. As every choreographer knows, while physically unrelated, in our perception, sound and movement are often blended together and even confused with one another. Each choice of mapping offers different directions of compliance. For example, 'more movement' may result in more sound, but it can also result in less sound. This may seem counter-intuitive, but there are actually situations where it feels exactly right. For example, holding a lifted shape, what dancers call a 'suspension', often require a good deal of energy from the standpoint of the performer even though very little movement is involved. Thus, this high energy situation, this moment of *less motion*, might be effectively mapped to *more sound.*

Let us look now at how qualities or parameters of a dance might be interpreted by a computer programme, the nature of the data which can be extracted, and finally how the data can be used to effectively influence our experience of a performance work (i.e. mapping in the parlance of dance tech). The image below shows a sample mapping for a scene in a hypothetical dance or theatre piece.

Must mapping be intuitive? No, but in practice it is difficult to achieve a meaningful result if it is not. Straying even a little from what seems intuitive in terms of mapping – what *makes sense* on a feeling level – will result in a piece for which the outsider loses all perception of inter-activity. And it is harder than you might think to get a cognitive or

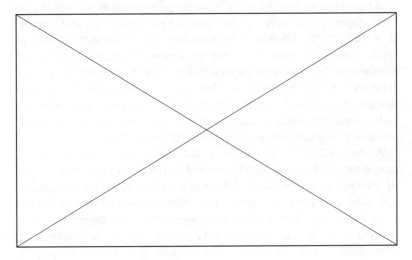

Figure 8 Multi-dimensional mapping

artistic effect from *any* mapping, let alone those that are complex, subtle, or obscure.

It has been argued on many occasions, by many artist/researchers (e.g. Richard Povall[10]), that the audience's ability to follow the mapping, that is to perceive the causal relationships of the media on stage, is not only unimportant but irrelevant. Since the technology has an influence on how the performer feels on stage, its use is justified because this will invariably affect the performance. The criterion, according to this viewpoint, for the successful application of technology in a performance is whether or not a beautiful work of art results.

This line of reasoning is troublesome for two reasons: One, because this justification is essentially a red herring. It simplifies the discussion beyond relevancy. The issue is not whether beautiful works of art can be made using technology, but rather if the same, or a very similar work could have been accomplished in a simpler, less distracting or pretentious way. The second reason it is troubling relates to the fact that in all areas of research (scientific and artistic alike) there is a tendency to read more into findings than they in fact contain. This is true wherever a degree of subjective judgment is involved and it is especially true where the researcher has a vested interest in a particular outcome.

In 1996, Beatrice and Allen Gardner taught a chimpanzee named Washoe to speak using American Sign Language (ASL). They claimed that the chimpanzees' difficulty in language acquisition was not due to their stupidity, but rather to an 'inability to control lips and tongue' (Patterson, 1978). Thousands of hours were spent teaching Washoe to speak. And, sure enough, he started speaking (signing). Articles appeared in scientific journals, newspapers, and television: 'The talking chimp' (visit www.friendsofwashoe.org). The Gardner's work was scrupulously documented through careful observation, videotapes, and extensive notes. Over the ensuing months and years, however, the Gardners' 'discovery' was completely discredited (Johnson, 1995).

Washoe did not learn to speak, but rather, like a dog, had been taught tricks. The scientists who worked on the project were not accused of manipulating their data, at least not *intentionally*. Noam Chomsky described their error as 'experimental delusion', adding that one needs to be extremely cautious to avoid 'suggestion by the experimenter' or a 'sentimental interpretation of data' (Chomsky, 1995). Despite their best efforts to maintain scientific impartiality, the Gardners were, quite simply, duped. They had become friends with their subject matter.

If nothing else, artists are friends with their own creations. Because audiences applaud, because well-wishers ooh and ahh, because newspapers write good reviews, artists are in no way excused of their responsibility to be sceptical when employing new technologies on stage.

Perhaps out of a desire to see something – *anything* – new, there is an almost mythical attraction to high technology in the arts. This is, of course, part of the fun of making art of this kind. But it is also dangerous. For creators, it leads to a 'boys with toys' mentality. The means overtake the end in importance. Like a magician, a 'how did they do that?' head-tilting in the audience arises. It is naive to assume viewers can watch works with new technology without at least a mild curiosity into the technical issues. To this extent, technology can dilute and obscure artistic intentions. To put it another way: there are stronger and weaker justifications for employing technology in a piece of art and the ability to analyse which are paramount to creating good work in this idiom.

Multi-dimensional mapping

Looking again at the sample mapping in Figure 8, one might think that such a complex array of interactive relationships might make an interesting piece. In fact, however, if all those things were happening at once it is not likely that you would perceive anything at all! Nice piece . . . but what were the computers for? There *are* situations where multiple mappings can be extremely effective, but in general it causes a diffusion which increases with the number of simultaneous mappings. This is not to say that complexity should necessarily be avoided. Palindrome made effective pieces which employed hundreds of EyeCon elements.[11] Mapping refers to *kinds* of casual relationships, not the complexity of any given system.

The pitfall of position tracking

New users of motion tracking systems such as EyeCon commonly have the idea to use it to track the position of an actor (dancer) in space. They wish to either define zones in the space which, when they are trespassed, produce different results, or they seek to define spatial tendencies: downstage=louder, upstage=quieter, and so on. Surprisingly perhaps, this is one of the least effective and most difficult implementations.

It is laden with pitfalls, both technical and artistic. From the technical side, the problems begin with the overhead position of the camera which such a plan would necessitate. Small theatres typically offer between 4 and 6 meters of ceiling height. Even with a relatively high 6-meter ceiling, there are considerable problems awaiting you. With a wide-angle (2.8 mm) or super-wide angle (2.6 mm) lens, it is possible to view the entire stage from this height. Coverage is not the problem. Rather it is that a camera placed 4–6 meters above a stage does not see the entire stage from above. This point is deceptive, but as the performer moves to one side, he or she is seen more and more the side, and less and less from above. Not only does this mean that the identifiable location of a player changes threefold or fourfold as they move around the stage, it also means that in the regions near the edge of the stage, one player can easily overlap another and the computer will see one person instead of two.

But even beyond these technical snafus, the approach is weak artistically. Between 1995 and 2002, Palindrome created six works based on this approach (Wechsler, 2005). Some, such as *Abstände* (1995) and *ManWoman Abständ* (2002), entailed continuous changes in the sound depending on the position of the dancer on the stage. In *Stolen (with respect)* (2000), the distance between two dancers in a duet controlled the loudness of the music (from a DJ). Additionally, the closer the dancers came to the corners of the stage, the louder various samples could be heard. The male dancer produced one set of samples, the female another (babies, anonymous telephone conversations, birds, street traffic, etc.).

Two further position tracking dances were made by Palindrome in which the distance between male dancers controlled the loudness of one track in the music, the distance between the females controlled a second and the distance between males and females controlled a third (Wechsler, 2005).[12] Since human beings control their distance between one another with great care, one might assume that this implementation would be effective. None of these works, however, met up to their expectations. This is not because they incorporated counter-intuitive mappings. They did not. As just described, the mappings would seem to make perfect sense. It is the approach *in general* which fails and does so consistently. The difficulties go beyond the technical ones described above. Position tracking appears to be weak for a variety of reasons: it lacks immediacy, it is not palpable or tactile, and finally location, as a parameter of movement, is simply of little interest to us compared to, say, body height, shape or movement dynamic. Let us look at these points separately.

A human body cannot immediately change its position on stage. It takes time, and this time frame defeats a convincing interactive effect. During the time it takes to change position, a musician or theatre technician might easily have simply moved a slider. Or, easier still, the performer could be following a cue in the music, what is utterly normal and expected on a stage. Thus, since the experience might have been created another way, the observer simply has no sense that anything special has taken place. The second reason that position tracking tends to fail is that it is rarely a physical experience. It is unlikely to be something that the performer knows in their tactile or kinaesthetic sense, and since it is not felt by the performer, it is not likely to be felt by the audience either.

Finally, location in a space is simply of little interest to us compared to parameters such as shape, acceleration, height, contact with the floor, contact to other persons, and so on.

Other relevant issues

Camera angles

If overhead camera positions are problematic, what *does* work? First of all, it is not the overhead angle *per se* which is problematic, but rather the use of it to map events which are spread continuously across a large area. There are many instances where overhead cameras are extremely useful. Palindrome has made a number of pieces recently in which granular synthesised sound samples are spread across the stage, making a kind of map or soundscape.[13] Because the musical compositions are highly varied, it is easy to feel the changes in sound even across minute distances. Perhaps the most useful camera position, however, is above, in front, and to the side (diagonal in all three senses of the word). It usually allows the greatest distance from the stage so that wide-angle lenses with their incumbent distortions can be avoided. This position has choreographic advantages as well. For one thing, it allows the detection of lateral, vertical, *and* upstage–downstage motion. It is, however, limited in that not *every* lateral, vertical and upstage–downstage motion can be detected (i.e. those directly towards or away from the camera). It also requires that the choreographer and performer both pay attention to the camera angle, as the perspective can be confusing. But if the artists pay attention to this point, the audience will not need to. That is, if you do it right, the illusion is created that every movement of the dancer *could* produce an effect, even if they are not being used at that moment. Obviously, this is a limitation, and additional cameras could solve it, but the argument here is that doing so may not be worth the trouble. Other problems (such as confusion!) will arise with additional cameras.

The interactive effect

If we accept the notion that transparency and intelligibility are important when using interactive systems, then the question follows: If a movement causes a sound or image, how exactly does the audience come to know this. That is, the fact that a movement and a sound have happened at the same time is not by itself at all unusual. Remember, dancers train hard to learn to synchronise their movements with sound. The interactive effect, then, clearly owes its efficacy not to synchronisation alone. What are these factors? What is it that gives us the sense that a particular body has a particular secondary medium under its control? And, moreover, how do different set-ups function, in the artistic sense, that is, in eye and ear and, as we have seen, mind of the observer? These are two separate questions: the first asks what makes interaction convincing, and the second asks what makes it interesting. Separating these is important. That is, the artist really has two separate tasks: one is to employ or design a system that works, or reads in the perception of the audience – i.e. is intelligible – and the other is to find how it can be interesting to an audience and useful to the needs of a performance piece – i.e. is artistic.

Artistry

Looking at the second question first, what is it about interactive art that makes it interesting, a number of answers come quickly to mind. They fall into three categories:

1. Amplification of gesture
2. Communication with an unseen player
3. Visual or acoustic accompaniment.

Amplification of gesture

Sensors, video cameras, or electrodes can make a tiny gesture have a large effect. In Palindrome's recent opera *Brother/Sister* (2005), the duet of the lovers contains small, intimate skin-on-skin contact between the performers. The gesture would normally mean more to the dancers than to someone sitting 20 feet away. However, with the help of an electrode-based system and live close-up video images, the effect of the contact is enormous. The touches are both heard in the music and seen as the contact moments also affects the video image in the form of a video image effect.

Communication with an unseen player

The best interactive performers we know are those with a sense of play. They have the attitude: 'If the machine is going to talk back to me, then I'll talk back to it.' This openness to realtime experiences and a willingness to respond to them in spite of being on stage makes the scene come alive and seem spontaneous and believable to an audience.

Visual or acoustic accompaniment

We and others working with such technologies used to say, instead of the dancer following the music, now the music follows the dancer. This, however, is actually beside the point and somewhat misleading. Live music, with its rich organic qualities and its live edge – the feeling that anything can happen – makes a fabulous accompaniment to dance. When it is well rehearsed, and when the musical and dance performers are in communication with one another, a lively and highly interactive feeling can arise. The beauty of that obviously does not lie in the fact that anyone *follows* anyone else. Control is not what interaction is about. Still, there is a sense in which live generation or control of music through movement mirrors live accompaniment. It generates singular moments which vary from performance to performance, and audiences sense this.

Intelligibility

Just as we have listed some things which make interaction artistic, we can likewise point to certain things which make it convincing. They are similarly dependent on both technology and psychology: Education, Timing, Repetition, Interaction by Implication and Intuitive Logic.

Education

Programme notes, a lobby exhibition, and interactive installation are milder forms of education than that of actually using the stage as an explanatory forum. Lecture demonstrations (either before or after the piece) are another approach. It obviously depends upon the context, the formality of the setting, and what kind of piece is involved. Some works do not need it, some cannot stand it, while others simply love it. Whether you do it or not, one thing is guaranteed: people will come to you after the show and thank you from the bottom of their hearts for it, just before the person beside them castigates you for the same thing. The question 'how did they do that?' is bound to arise. Human beings are naturally curious. It is neither realistic to deny this common instinct, nor is it necessary to turn every theatre piece into a science class.

Timing

Audiences seem to have a tolerance of about half a second latency in interactive pieces. That is, if more than this time elapses between motion and generated media, both the performer and the observer will lose the sense that the first event is causing or affecting the second. Indeed, even that half second can be quite annoying.

Repetition

When a movement triggers or changes a sound or image, how do we know that it is not coincidence? Or may be the scene was simply well rehearsed? If the event happens only once, then we really have no way of knowing. Thus repeating the event, at least in early instances, is highly recommended. Exactly when, where, how, and, in a theatrical sense, why this repetition takes place is of considerable importance, which brings us back to the point about performer's role in all of this. If the performer is keenly aware of their environment and the effect they are having on it, then the audience probably will be as well. Performers who are able to be relaxed on stage, and in this sense vulnerable, are often the best in this type of work.

Interaction by implication

When some interactive moment, not too far into the piece, is made utterly clear and convincing, then audiences will cut the piece a certain amount of slack, so to speak, and will find themselves sensitised to subtler mappings as well.

Intuitive logic

As we have discussed, there are more and less effective parameters to employ in any given artistic situation and the choice is of considerable importance. There is a tendency for artists to employ mappings and systems which are more complex than necessary. There is a tendency likewise to want to try out ineffective counter-intuitive mapping and compliance.

Conclusion

The special performance qualities which can be achieved through interactive technology are not well explored. While we are beginning to learn what they are and how to control them, they remain in other cases illusive and highly subjective. The fascination with the means can confuse the end. Technical issues such as tracking accuracy, resolution, noise, and interference are often confused with efficacy from an artistic standpoint. A digital system which seems like the future now will bore the entire culture a few years hence. Systems written off as passé or gimmicky re-emerge in the next instance used by an artist to surprising effect. We have to conclude that the central challenge that this field faces is not one of improving the technology, but rather one of developing an understanding of its implications – the changes in the mindset and sensibility of artists as they put it to use.

Notes

1. The use of mechanical devices such as the *ekkyklema* ('a wheeled-out thing') and the *mechane* ('theatrical machine') are well documented, though the details concerning how they worked are unknown (Dunkle, 2006).
2. While anecdotal, this would not be out of character. King Louis XIV encouraged an extraordinary blossoming of culture; theatre (Molière and Racine), music (Lully), architecture, painting, sculpture, mechanical engineering (Camus) and all the sciences (founding of the royal French academies). He was known to have been a performer himself and was particularly intrigued by mechanical toys and remote-acting devices including weaponry, such as rifles, and canons, which were developed during his rein (Histoire-en-Ligne, 2005; Steingrad, 2005).
3. 'Rauschenberg's performance works incorporated the everyday and the unexpected; dancers often used ordinary, untrained movements, while stage sets were sometimes composed of found objects, Combines, and live decor in which human activity and stage props were indistinguishable' (Blaut, 2005).
4. *When summoned*, 1969, music by Morton Subotnick (Evans, 2005).
5. Founded in 1994, Troika Ranch is a digital dance theatre company which has the mission to create live performances that hybridise dance, theatre, and interactive digital media. Based in New York City, Troika Ranch is the collaborative vision of composer/media artist Mark Coniglio and choreographer Dawn Stoppiello (Coniglio and Stoppiello, 2005).

6. Ventura Dance Company is based in Zürich and is lead by Pablo Ventura. They work with emmersive projection environments, robots, EyeCon motion tracking, and other computer-based systems (Ventura, 2006).
7. Palindrome was founded in 1982, by this chapter's author. In 1995 computer engineer Frieder Weiß joined the company as has been its co-director since 1998. Palindrome has largely been focused on the application of motion tracking technology to dance, though they have also been responsible for a number of installation art works as well (Wechsler, 2003).
8. The largest project to date is the motion project of the Arts, Media and Engineering (AME) programme at ASU, which brought together choreographers Trisha Brown and Bill T. Jones; media artists Paul Kaiser, Shelley Eshkar, and Marc Downie; composers Roger Reynolds and Curtis Bahn; lighting designer Robert Weirzel with AME artists and engineers for the creation of two interactive multimedia works, new motion analysis systems, and interactive technologies. Thanassis Rikakis and Colleen Jennings-Roggensack were project co-directors (Motione, 2005). Other work in this direction has been undertaken by Luc Vanier *et al.* at the interactive performance facility at University of Wisconsin at Milwaukee (Vanier *et al.*, 2003).
9. Kalypso is a realtime video processing software, used by Palindrome and others, though it has not yet been formally released. For more information, see http://eyecon.palindrome.de (Weiß, 2005).
10. For example in Dance and Technology Zone email discussions: http://www.scottsutherland.com/dancetechnology/archive/2003/0193, accessed 6 January 2006; http://www.scottsutherland.com/dancetechnology/archive/2003/0215, accessed 6 January 2006; http://dancetechnology.com/dancetechnology/archive/2000/0312, accessed 6 January 2006.
11. *Minotaur*, 2000, was composed by Erling Wold (2005) and was performed extensively in Germany and Austria during the years that followed.
12. *With Sounds of Children*, 1999; and *Quartet*, 2002 (Wechsler, 2005).
13. A video is available at http://www.palindrome.de (Wechsler, 2003) under 'new works and works in progress'.

References

Blaut, J. (2005). *Caste: Exposicion: Robert Rauschenberg: Contenido.* http://www.guggenheim-bilbao.es/ingles/exposiciones/rauschenberg/contenido.htm, accessed 6 January 2006.

Chomsky, N. (1995). Recording and transcript by Radio Free Maine. University of New Hampshire (April 12).

Coniglio, M. (2006) *Troika Tronix.* http://www.troikatronix.com/isadora.html, accessed 6 January 2006.

Coniglio, M. and Stoppiello, D. (2005). *Troika Ranch.* http://www.troikaranch.org/company.html, accessed 6 January 2006.

Dunkle, R. (2006). *Introduction to Greek and Roman Comedy.* http://depthome.brooklyn.cuny.edu/classics/dunkle/comedy/index, accessed 15 February 2006.

Evans, B. (2005). *Bill Evans Danc.* http://www.billevansdance.org/, accessed 6 January 2006.

Histoire-en-Ligne. (2005). *Louis XIV le Grand dit le Roi Soleil.* http://www.histoire-en-ligne.com/article.php3?id_article=225, accessed 6 January 2006.

Johnson, G. (1995) 'Chimp Talk Debate: Is It Really Language?' *New York Times* (June 6).

Maya. (2006). *Animation Arena: Art, Models, Articles.* http://www.animationarena.com, accessed 6 January 2006.

Motione. (2005). *Motione.* http://ame.asu.edu/motione, accessed 6 January 2006.

Patterson, F. 1978. 'Conversations with a Gorilla'. *National Geographic* (October).

Povall, R. (2003). *What was Missing?* http://www.scottsutherland.com/dancetechnology/archive/2003/0193 and http://www.scottsutherland.com/dancetechnology/archive/2003/0215, accessed 6 January 2006.

Steingrad, E. (2005). *Louis XIV The Sun King.* http://www.louis-xiv.de/louisold/louisxiv.html, accessed 15 February 2006.

Vanier, L. *et al.* (2003). *Connecting the Dots: The Dissection of a Live Optical Motion Capture Animation Dance Performance.* http://www.isl.uiuc.edu/Publications/final%20dance1.pdf, accessed 6 January 2006.

Ventura, P. (2006). *Ventura Dance.* http://www.ventura-dance.com, accessed 6 January 2006.

Vicon. (2006). Vicon website. http://www.vicon.com/products/systems.html, accessed 6 January 2006.

Wechsler, R. (2003). *Palindrome Intermedia Performance Group.* http://www.palindrome.de, accessed 6 January 2006.

Wechsler, R. (2005). *Palindrome History.* http://www.palindrome.de/history.htm, accessed 6 January 2006.

Weiß, F. (2002). Radio interview in *Kulturheute*, Bayrischen Rundfunk, München (3 April).

Weiß, F. (2005). *EyeCon.* http://eyecon.palindrome.de, accessed 6 January 2006.

Wold, E. (2005). *Erling Wold Composer.* http://www.erlingwold.com, accessed 6 January 2006.

6
Materials vs Content in Digitally Mediated Performance

Mark Coniglio

After two evenings of performance at the Digital Cultures Lab (2005) in Nottingham, England, several colleagues gathered at the local pub to discuss what we had seen. An interesting thread emerged as we debated the various works, which included a work-in-progress version of Troika Ranch's *16 [R]evolutions* (2005): must media-intensive dance performances focus, throughout a work's duration, on fully exploring and/or revealing a single interactive system and its visual or aural result? Or, is it feasible to have highly varied visual/aural results as long as they rigorously support the aesthetic intent of the piece? And, what does the difference in these two modes of exploration tell us about mediated performance work in general?

To give an example of this dialectic, I want to give brief impressions of the approach used in two mediated works: *Apparitions* by Klaus Obermaier (2004), and *16 [R]evolutions* (2006), created by myself and choreographer Dawn Stoppiello. Both works use motion-tracking systems to interactively generate three-dimensional visuals that respond to movements of the dancers, and both rely on a unified methodology for using those tracked movements to manipulate the media. In *Apparitions*, the visuals stay within a clearly defined realm throughout the course of the piece, and these visuals are visible throughout almost the entire work. *16 [R]evolutions*, on the other hand, presents a far more varied palette of visual imagery that is present in several sections, but is, significantly, absent in others. And, in a few spots, the imagery is not interactively controlled at all.

A work like *Apparitions* follows a clear tradition in art making: exhaustively explore the attributes of all materials used in a work. This exhaustive approach is apparent in everything from compositions, dating from the early 1700s, for the newly invented technology of the

Figure 9 The vertical line establishes the first graphic theme in *16 [R]evolutions* (Performer: Daniel Suominen – Photo: Richard Termine)

pianoforte to Pierre Schaeffer's *musique concrete* manipulations of the sounds of the Paris Metro (1948) or James Joyce's *Finnegan's Wake* (1959). And while I make no argument against the rigor applied in these explorations, I am drawn to notice that these works are, in essence, *about* the materials themselves.

In contrast, during the Digital Cultures Lab, the work of Troika Ranch was described in a paper by Isabel Valverde (2005) in a way I had not previously heard, but that seemed quite appropriate to me: *content-driven*. Our approach has always been to use dance, theater and media to create contexts and express tensions focused on a single concept or idea. And it is specifically our interest in content that marks a divide between our work and the aforementioned explorations that focus on fully exploring materials. Because we are interested in a theater of ideas, we attempt to join the visual imagery with the dance and theatrical sections in support of these ideas. The rigor comes in carefully relating all materials to the core narrative theme of the work.

Still, during our discussion, one colleague contended that, since I was not dedicating myself to a full exploration of the digital materials and the performers' interactions with them, the work lacked rigor. John Rockwell's review of the full version of *16 [R]evolutions* in the New York Times makes perhaps a similar argument when it says 'Mr. Coniglio and Ms. Stoppiello apparently don't have faith that their technology will provide enough variety or meaning to sustain interest over an hour' (Rockwell, 2006). My response is that full exploration of the narrative arc is of primary importance, and that the materials, whether they be movement, text, or visuals, can be brought to bear as needed as long as they support the aesthetic intent of the piece, and as long as they do not appear as *deus es machina*. (Certainly the latter issue was present in the version of *16 [R]evolutions* shown at Digital Cultures – the final version premiered in New York in January 2006 resolved these issues – in part as a response to the useful critique offered by our colleagues.)

To support this, I pointed to a work of Pina Bausch (1997), *Der Fensterputzer*. In this work, a gigantic, three-meter high pile of red bauhinia flowers is present on stage. The importance of this object was clear, if only gauged by scale. But only occasionally do the performers directly refer to this object – mostly it is just *there*. Certainly it was never 'fully explored' in the sense that the dancers manipulate it in every imaginable way. Nor was its meaning was ever made explicit. But, simply by being present on the stage, it successfully created a context that changed the way we viewed everything happening around it, as well as linking every section of the piece by its presence. Why might not the media in a mediated performance be used successfully in the same way Ms Bausch used this mountain of flowers?

A certain importance is also placed on the fact that all of us in the dance-tech arena are using 'new' technology, and that its recent addition to the vocabulary begs an exhaustive exploration of its possibilities. Using new technology to further expression has always been the realm of the artist, but how often have these explorations that focused on the technology itself remained important in our canon? While the early films of the Brothers Lumère are important, they do not have the resonance of *Citizen Kane* (Welles, 1941); the same holds true for the early *pianoforte* works when compared to the later compositions of Chopin. And, while it may be true that we cannot have *Citizen Kane* without

Figure 10 Traces of the performer's hands and feet leave multiple curved white traces, a development of the white line seen earlier in *16 [R]evolutions* (Performer: Lucia Tong – Photo: Richard Termine)

A Trip to the Moon (Melies, 1902), I feel we need to question where we are in the progression of the new technologies as applied to mediated performance.

One problem in this regard is that new technologies are constantly appearing, so how can one develop upon the previous body of work and search for the depth of meaning that comes with an evolutionary use of the technology? For example, how many recent performance works explore the technology of the electric light as an idea in of itself? Save a few contemporary examples (Seth Riskin at IDAT '99 comes to mind [Riskin, 1999]) one has to travel back many years to find such explorations. Instead, theatrical lighting technology has developed to the point where it is most often used to support the narrative mood of a performance, and its presence as a technology is not questioned *a priori*. The electric light is so integrated into our theatrical (and societal) experience that exhaustive exploration of it seems, generally speaking, unnecessary. Are we at the Thomas Edison stage of dance-technology, or somewhere further down the line?

As I try to somehow give a name to what might be described as the *materials-driven* and *content-driven* camps, I keep returning to the terms *abstraction* and *narrative*. Or, perhaps, if one wanted to link the camps to a métier, one could say dance vs theater (dance used here in its more

Figure 11 Color and multiplicity are introduced to imply evolutionary change in *16 [R]evolutions* (Performers: Johanna Levy and Daniel Suominen – Photo: A.T. Schaeffer)

contemporary, abstracted sense; theater understood by its more traditional literary interpretation). To focus on fully exploring the materials means that you are making those materials the subject of the piece – which arguably falls under a well-established (reactionary?) modernist agenda. To use the materials to suggest, or provide context for, an idea implies narrative – put simply, someone is telling a story. This would most certainly not fit under the modernist directive, though it too could be reactionary too, as storytelling is certainly quite a traditional mode of presentation.

What does this tell us about mediated work in general? Primarily I see that there is a serious compositional problem when attempting to combine technologically advanced visual imagery with content-based performance, because that imagery (especially shown in a large scale) is inherently seductive in a way that is quite difficult to control; admittedly, it is a very different animal than Ms Bausch's carnations. In *16 [R]evolutions, Apparitions*, or even Carol Brown's lovely *See, Unsee* (2005) work shown at Digital Cultures, the imagery is so beautiful that it borders on the erotic. Referring again to the New York Times review, Mr Rockwell is quite correct: I have absolutely no faith that the technology (the visuals and interaction) will provide enough meaning to communicate the themes in which we are interested. Instead I propose to rely on all of the elements to intensify each other in a complementary manner, so that the viewer's attention remains intensely focused on the content of the work. His statement tells me either he was not rigorous as a viewer or that we failed in contextualizing and relating the materials placed before him, that is that we let him be seduced by the visuals to the point where the other materials became invisible. If the latter is the case, then something has gone wrong within the compositional process on our part. But if it is the former, it may not be his (or any viewer's) fault. It could be that the newness and novelty of the imagery itself makes it difficult (or impossible) for the more inexperienced viewer to focus on anything else.

How do we move past this stage? Certainly, if the problems described above exist, it is an argument for the exhaustive approach, which follows the historical model by proposing that we need to experience work focused on technique and methodology before we can move on to a deeper exploration of more human (i.e. content-driven) themes.

But, because it is the function of the artist to bring new technologies to the table as rapidly as possible, it seems entirely problematic for content-driven creators to wait because there is no end in sight to the new imagery and interactive possibilities offered by new technology. In the end, my proposal is that those creating content-driven work must forge ahead in parallel with those who are creating materials-driven work. Certainly some of these works will falter; we will see materials-driven pieces that do not speak to our human experience, content-driven work that fails because we cannot integrate the imagery with the other métiers of the piece. But, if these two camps pay careful attention to each other as we present our work, and audiences are able to experience both modes, we have the opportunity to go beyond the historical model as we holistically and simultaneously develop theory, technique, and content.

References

16 [R]evolutions (Work-in-Progress). (2005). Direction, Concept: Mark Coniglio and Dawn Stoppiello; Choreography: Dawn Stoppiello; Music, Interactive Media: Mark Coniglio; Dramaturgy: Peter Salis. Digital Cultures Lab, Sanford Theater, Nottingham, England. 2 December.

16 [R]evolutions (Premiere). (2006). Direction, Concept: Mark Coniglio and Dawn Stoppiello. Eyebeam Art & Technology Center, New York, New York, USA. 18–28 January.

Bausch, Pina. (1997). *Der Fensterputzer*. Direction and Choreography: Pina Bausch. Next Wave Festival, Brooklyn Academy of Music, Brooklyn, New York, USA. October.

Brown, Carol. (2005). *See, Unsee*. Concept Carol Brown and Mette Ramsgard Thomsen. Choreography: Carol Brown. Architecture and Interactive Media: Mette Ramsgard Thomsen. Sound: Alastair MacDonald. Performed by Carol Brown and Katsura Isobe. Digital Cultures Lab, Sandford Theater, Nottingham, England. 3 December.

Digital Cultures Lab. (2005). Coordinator: Johannes Birringer. Nottingham-Trent University, Nottingham, England. 28 November–4 December.

Joyce, James. (1959). *Finnegans Wake*. New York: Viking Compass.

Melies, Georges. (1902). Director: *Le Voyage dans la Lune (A Trip to the Moon)* Paris: France.

Obermaier, Klaus. (2004). Concept, Idea, Direction, Visuals, Music. *Apparitions*. Ars Electronica, Posthof Theater, Linz, Austria. 4 September.

Riskin, Seth. (1999). *Light Dance*. Created and Performed by Seth Riskin. Galvin Playhouse, International Dance and Technology Festival (IDAT) '99. Arizona State University, Tempe, Arizona, USA. 26 February.

Rockwell, John. (2006). *Tracking Mankind's Rise, With the Help of Computers*. New York Times. 20 January.

Schaeffer, Pierre. (1948). Composer. *Etude aux chemans de fer*. Paris, France.

Valverde, Isabel. (2005). Unpublished paper.

Welles, Orson. (1941). Director. *Citizen Kane*. RKO Pictures, California, USA.

7

Learning to Dance with Angelfish: Choreographic Encounters Between Virtuality and Reality

Carol Brown

> In searching for the something else beyond self and other, what or who are the we that haunts us? Who are the strangers at the heart of the self who disrupt our sanctuary with disquieting moments? Mothers, dogs, sea urchins, whores, mystics, muggers, diseased spores, derelictions and secretions. I spawned multitudinous becomings within a constantly deformable body, a malleable container of anarchic desires. I became so lucid that I, in becoming not-I could disappear beyond a thousand species of diverse others. My fantasy was to be everywhere and no-one. To cast off this sluggish flesh and become hyperreal. To glide with sea creatures in a rock pool phantasmagoria.

Through digitally extended performances bone memories mix with machine memories fusing the gravitational flows of the dancer in space-time with the place-unboundedness of digital forms. In learning to dance with data, spaces unfold, striating the present moment through multiple dimensions. In these matrixial spaces the stage metamorphoses from a physical location – grounded, fixed, actual – to a relational space – incorporating the ungrounded, the fluid and the virtual. Performance identities which were previously place-bound become mobilised and de-territorialised. The privileged state of performance as a 'being here' in the elusive present is no longer embodied in the taken-for-granted 'thereness' of the stage of soil and flesh, it becomes a superabundance of becomings experienced as hyper-realities and distributed presence. In this context, the 'being in the body' of embodiment is radically reconfigured.

Learning to dance with Angelfish is a metaphor for the experience of re-learning embodiment through live interaction with a virtual dance partner, a creature of code who slips from my grasp and swims at the edges of my vision. In this writing, I am reflecting on recent experience in the field of interactive performance through a critical prism which addresses the shifting 'nature' of dance and the (dis)embodied subjectivities of dancers dancing in the digital realm. The page becomes an interface for intersections of thought and practice, exposing the 'second nature' of the dancer whose body thinks itself by probing the space of the present.

We start from where we are and we build on what we know. Dancing, as the articulation of movement in space and through time, has historically taken place within three dimensions but the space of the present is both actual and virtual *in nature*. As Paul Virilio explains, there are today two inter-related spaces: 'next to actual space, which has been the space of history, there is now virtual space, and the two are interdependent' (2002: 67–68). Through dancing space unfolds. In the refolded space of data dance we discover a haunting virtuality and a new biodiversity of material-informational figures.

Brian Massumi citing Giordano Bruno describes the virtual as the ' "real but abstract" incorporeality of the body' (2002: 21). Given that the virtual is a force which acts in another dimension, as 'a continuous unfolding on the road to becoming other', the primary challenge for choreography in mixed reality environments is to create 'new movements toward the virtual by tripping up repetition, purging habit and reason, and encouraging difference' (Beckman, 1998: 16). Choreography, as a *writing* of spaces through the moving body, embraces this challenge through merging layers, intercutting between dimensions, dancing thresholds and streaming visceral thinking with 'travels in virtuality' (Thomas, 2004).

Spawning identities

So, shall I tell you how we made it? How my cellular and its data hatched a different kind of being and how this being met us with its machine eyes, but only in outline, it missed all the inner lines – the creases of our gestures, the movement of our eyes, the tone of our touch – it made a photofit of our nerve endings and grafted this onto its own skin like virtual tattoo. We trained it to track us like a distant geography, never getting too close, never getting the scent of this skin at close range. It followed our contours until it learnt to predict our futures, where we would move to next. It did not imitate, it created, growing children in its body like fish roe. It changed. We made fine calibrations inserting new memories in the iliac crest of the pelvis, in the mastoid bone of the skull, in the cervical vertebrae. We communicated through the outerspheres of each other. We made insertions and we learnt to touch that which we could never hold.[1]

I am learning how to move within a technological habitat with a digital infrastructure. Through the creation of an embodied interface – *Spawn* – in collaboration with architect, Mette Ramsgard Thomsen, we are conceiving embodied interfaces as tools for the creation of performance spaces which integrate fragments of reality, virtuality and fantasy. *Spawn* is an interactive 'stage' informed by a camera-based interface. Dancers negotiate a jointed space, moving between tracked and non-tracked zones of the stage and calibrating their experiences of these differentiated spaces. In the tracked zone, a machine vision system, using a single side-mounted camera, identifies the shifting outlines of the dancers' bodies. Rather than identifying body parts and tracking their movement in two- or three-dimensional space, the *Spawn* interface generates a set of statistical characteristics of the dancer's silhouette size and shape. This data becomes input for a *virtual other*, a digital morphology shaped by the presence and movement of the performers.[2] The *virtual other* is a complex geometry comprising four circles stretching a spline-based membrane between them. As the performers move, they affect the *virtual other* deforming and reforming, contracting and expanding, folding and unfolding its digital skin. The visualisation of this kinetically modelled avatar is projected back into the physical space of the performance in real time, generating new forms of interaction and creating a blended environment of real and virtual spaces for a mobile audience.

Unlike other computer interfaces used in dance, such as Hypervision MoCap, the *Spawn* interface does not seek to identify the dancers' body parts and map them onto a corresponding digital anatomy, mimeticising the morphologic of the dancer. Instead, a set of statistical characteristics of the dancer's silhouette size and shape are generated and updated in real time. The digital is conceived as a separate dimension, informed by the embodied presence of the performer, yet retaining an independent morphology and motility.

They set out, like two explorers with borderline personalities in a hallucinatory room. They ascended without abandoning the earth; they awakened energy without capturing it. The contours of their movements were tracked in another dimension, but the inner lines of their postures and intricate complexions of their gestures escaped the seen/scene. From their outlines statistical data refashioned them in dialogue with a 'sphery thing', a shivering architecture which would never stand up. Their place became the crisscrossing of spaces, a threshold between the actual and the virtual requiring a simultaneity of perception. Because we make a home for ourselves wherever we happen to be, in this virtual and actual habitat they experience a life together.

Given its radical difference from our own bodies and spatial histories, the rehearsal process for *Spawn* involved a reconfiguring of movement

Figure 12 Catherine Bennett in *The Changing Room* (Photo by Mattias Ek)

to enable effective communication with and through the *virtual other*. In this process we came to know and relate to it, attributing anthropomorphic characteristics to its multiple appearances, and allowing energies to emerge through interaction between the different states it inspired. As we grew to know our virtual dance partner, our naming of it shifted from the 'sphery thing' to 'It'. This 'It' inspired a diversity of images as molecular, planetary, celestial and aquatic forms. It became an Angelfish, a Mollusc, our Virtual Puppet, an Infinite Cage, an Irish Sun, Iris and BloodMusic. Through our postural morphing, form enfolded form; contour wrapped and invaginated contour, space swallowed space. We hinged and flexed around and through each other creating a recombinant morphology.

> Extreme Width Curdling
> Extruding Warped Rotations
> Invading Walking Deformations
> With Rippling Splines
> Head Threading and Knees Spanning
> Diaphragm Oscillating and Vibrating
> Volumetric Hollowing
> Moving Underside Undermind
> Lines Inside and Outside
> Folding Back Into
> Grafting Impossible Anatomies
> Nurbs With Nervous Architectures.

Morphology is understood in this context, not as a purely anatomical image or model but as that which shapes understanding between forms, and between body, language and subjectivity. In *Spawn*, a morphological mediation occurs between the actual bodies of the three women dancers and the fleshless figure of the virtual object. Identity figures virtually in this 'through-otherness'.[3] Through an open relationship between corporeality and virtuality, two distinct spatial narratives, the one, somatically informed embodiment, and the other, spline-based geometry of a digital architecture, overlap and their surface areas are brought into contact through remote touch. Rather than harness the *virtual other* with its impossible geometries into our existing systems and making it speak as 'other of the same', we negotiated the identity of our performance through improvisational processes that incorporated its geometries and spatial logics.[4] This involved moving between a geometric or bilateral body symmetry based on the cross (horizontal–vertical axis and anterior–posterior axis) and a biometric or anthrometric model based on radial symmetry (concentric circles which move from centre to periphery and periphery to centre). Whereas the former relates to the geometry of the Western picture plane with its Cartesian coordinates, the latter is derived from experiential anatomy and the core-distal connectivity of developmental movement.[5] In working between these two models we had a device for negotiating embodied relationships between different conceptions of space, in particular between the two-dimensional tracking of the dancer's silhouette as an outline, and its mapping onto the possible 11 dimensions of the *virtual other*. As tools for improvising these different schemata of bodily organisation opened the possibility for a mediation of different morphologies and a relearning of embodiment.[6]

Virtual stages

The choreographic process provided experimental space and time for the collaborative team of dancers, choreographer, architect and programmers to encode and embody the potential of the virtual architecture, our *sphery thing*, however the work itself was incomplete without an audience.[7] In considering the audience's role and experience of the work we confronted the history of the western stage and the customary critical distance and fixed position of the audience. Dance, as a live theatre art, has inherited the scenographic conventions of perspectival vision through front-end stages, proscenium arch theatres and black box spaces. Historically, gravity as a primary structure for dance is expressed in figure/ground relations within the picture plane of these spaces. At its most primal level, this is about the relations and interconnections between body, stage and earth. However, when we add the virtual into this mix we are metamorphosing relations within and between these elements. The groundlessness of the digital requires a reconceiving of the relationship between figure and ground to enable what Virilio describes as our contemporary 'stereo reality' to be experienced by audience and performer alike (2002: 68). Previous works, such as *Shelf Life* (1998), explored this tension by siting the dancer, myself, on an elevated plane, off the ground and on a transparent 'stage' where her movements matched and counterpointed the gravity-free fluidity of her digital dance shadow.[8] Audiences experienced this work within live art settings where they could enter and leave at any stage during the 4-hour durational performance installation and view the work from any angle, including its underside.

Gravity is a primary structure for the dancer whose movements express a dynamic relationship with its forces. The body visibly expresses the gravitational forces acting upon it but it equally responds to the appearance of gravity-free digital environments. In inviting non-specialist movers into the interactive space of performance to interact with the virtual image, most will intuitively lighten their body effort and weight. We adjust our patterns of behaviour and actions in response to and in anticipation of changes to the patterns of activity around us. In dancing within digital environments our kinaesthetic and proprioceptive senses shift from the axes of a gravitational force field to the matrices of interconnectivity. Movement becomes less focused on the vertical axis as the dancer reaches and extends her limbs in the horizontal plane to 'touch' through the *virtual other*. The dancing body becomes a transducer, manifesting the rhythms, flows of energy, gravities and sensations of the interactive experience. Dancing, which makes connections with a weightless environment, heightens perceptions of the body's weight and potential for weightlessness, but it also leans perceptual awareness towards understanding space as curved and gravity as a by-product of the shaping of space itself.[9] In this context, surface tensions figure as much as gravitational flows in creating the underlying structure for the dance.

Furniture for Angelfish

Heightening the surface tensions of the dance over its gravitational axes has the effect of suggesting other ways of being a body. This brings the dancing subject into closer alignments, and potentially affinities, with other species whose morphologies and unique characteristics prescribe that they experience space differently. The Angelfish of the title of this chapter suggests this metamorphic potential. As the dancer learns how to move with her virtual dance partner, she is coming to know that which is beyond or outside her biological history. Deleuze has conceived of a posthumanist trans-species 'becoming-animal, becoming-plant' as a way of defining the self not through identity but multiplicities. A multiplicity is defined not by its centre but by its outer limits and borders where it enters into relations with other multiplicities and changes nature. In this context, the self is understood as a threshold between multiplicities (Smith, 1997). One can enter a zone of becoming with anything – plants, animals, minerals, fish – provided you find a method for doing so.[10] For Deleuze, these others are not about metaphors but are rather about metamorphoses. Stories of metamorphosis draw much of their power from their ability to override the mind–matter distinction and provide new ways of telling the self. The feminist Deleuzian Rosi Braidotti describes how the 'morphological hybridity' of metamorphic processes transgresses and erases bodily boundaries (2002: 128). Animals, and in this context fish, thus become images for orienting oneself in a strange territory through a processual metamorphoses creatively combining figures of thought from technology, zoology and biology.

Spawn is intended as a digital scenographic invention – an interactive stage – designed to challenge the duality of figure/ground relations and provoke a different kind of agency from the conventions of twentieth-century performance. Through a set of metamorphosing relations that allow a state of flux to exist between the real and the virtual, the uncontainable and the contained, we leave behind the territory we know and we enter a through-place. This place is in process and fluid, it is a place of passage where relations are fleeting and formed through contingent crossings of thresholds which resist fixed identities.

The performance system *Spawn* evolved into the event, *The Changing Room*. *The Room* was a nine-metre square performance area divided by long curtains which were manipulated by the dancers. The audience was guided through the performance to encounter each stage of the work from a different point of view, a different side of the room. They looked into the room with its furnishings and its embedded screens and they experienced not just that which is going on in front of them but also behind and around them as they negotiated their encounter of the work as a physical journey.

Part dance partner and part extended architecture, the three women performers experienced their changing room through a series of transformations: A mirror became a screen for their mutations; a curtain a technological frontier; and their table a platform for the puppetry of the virtual. In replaying traces of otherness embedded in their own memories, they explore the unfamiliar and the strange. Moving at the threshold between different dimensions of space, their gestures are tracked in the virtual environment. Embedded within the furniture of their room are a series of screens through which their virtual dance partner is rendered mirroring, extending and distorting their behaviour.

The everyday transformations of appearance which we experience in changing our clothes became a metaphor for the mutating forms inherent in the choreography. Through the mutable qualities of the room, the status of the everyday is extended, altered and augmented.

Familiar cultural objects become transformative moments, enabling the transgression of presence into the extended environment of the digital.

The choreography enabled a series of contingent and oscillating relationships between the performers and the virtual object to emerge. The detachment of gesture through the tracking system invokes an extended presence that is communicated to the audience through kinaesthetic perceptions, sound and vision. This extended presence shifts the centre of gravity beyond the primacy of the performer as the focus within the staged event, redistributing it between surface projections, malleable furniture and virtual object. Similarly for the dancers, their attention shifts and alternates between live and virtual presences as they respond and project the sensations within the room and communicate these to the audience creating a triangulated circuit of interactions. This effect is further amplified by mobilising the audience. Performing the role of 'host' and interlocutor, I invited the audience to 'evolve' with the performance by changing their point of view moving around the stage to experience the work from three different facings. Shifts in the audience's spatial relationship to the performance opened up its potential readings, allowing meanings to be uncovered in the interconnections between media, spaces and bodies.

In working with embodied interfaces we have a potentially powerful tool to dismantle the stabilities of the unitary body-subject, revealing the fissured and multiply stranded alignments of the dancer whose 'being here' is an unfolding of many differences. In this context, identity is reconfigured through choreographies that create new belongings. If, as Brian Massumi explains, 'I' am a being in becoming, then through the unfolding of experience, identity is not concrete or fully formed but emerging. In the situation of dancing the virtual, I am a being-in-process prior to that which can be named, marked and branded. There is a liveliness here because we do not know what we are in the process of becoming, we are only making discoveries and producing spaces, not in our likeness but through a metamorphosing of relations between different ways of being and in response to the bio- and techno-diversity of the world in which we live.

Choreographically, working with this conception of virtuality ruptures classical conceptions of space and time because it does not assimilate the virtual into our own self-image, rather it acknowledges difference through the co-presence of different kinds of inhabitation. This is not easy. With the embodied interface *Spawn* movements in the tracked space are co-opted, mediating relations between the *virtual other* and the overall choreography. This required the dancers to relate as strongly to the visual image as much as to the soundscape and to each other as dancers. Vision is our most objectifying and distancing sense, in relating to her virtual dance partner through screen projections the dancer takes her bearings from an external reference point. Moving in this way can be experienced as alienating and, at times, disembodying because it refuses the customary focused attention of the dancer, diffusing her agency through multiple layers of performance. The dancer needs to simultaneously incorporate the alternative morphology of the virtual and excorporate the multi-focused vectors of its performance through a redistributed agency. Re-learning embodiment through the experience of interactive dance in this way may be experienced as fragmenting, unsettling and destabilising. Perhaps because of this the development of the work from its research phase, towards its production phase, required the generating of 'stories' which narrativised and gave meaning and thickness to the stage space and the metamorphosing relations between dancer and *virtual other*.

Stories emerged through the research process as the dancers came to develop a *feeling for the avatar*.[11] Stories are one way through which we find our orientation in the world; they provide locative maps for navigating the unfamiliar and for making connections between fragments of reality, virtuality and fantasy. In *The Changing Room* we overlaid a virtual reality onto an everyday reality. In leaving behind the territory in which we were, we found ourselves in another place, beyond what we already knew. We slipped in and out of clothes which would morph our presence, extending and extruding lines between us, elasticising spaces, giving a sense of time and space beyond the scale of our own lives. We touched, hacked, mimicked, enveloped and became enveloped by the *virtual other*, shifting states and making up stories which bound us into a shared sense of presence, a shared sense of play.[12]

The internal contradictions and deep crevices of identity are not resolved by work in digital environments but rather made manifest. The incorporation of the *virtual other* through the inhabiting of unfamiliar dimensions, the redistribution of power in live performance through the mapping of the physical into the digital, and the sense of touching and enfolding the immaterial inform how and why we move, spawning new identities. As Miwon Kwon (2000) explains, an intensification of spatial departicularisation might exacerbate the effects of alienation and fragmentation through a loss of identity, but it might also provide a space for the retrieval of lost differences for, as Henri Lefebvre states, 'a new space cannot be born (produced) unless it accentuates differences'

(1991: 52). This might become more than a performance experiment as it is also a potential life strategy towards a more pluralistic understanding of the self.

Outmoded constructs of identity continue to wound with a territorialism which at its worst fosters a nationalism which claims exclusive possession of places to the exclusion of the Other, the Stranger, the Alien. In learning to love the alien, the stranger and the avatar, we are dancing differently.

To be sure, we will perhaps discover in foreign lands traces of gods that we are lacking. But, without a journey in ourselves, to celebrate with them will not really be possible. Approaching gods is not limited to discovering that they exist. It is in the intimate of ourselves that a dwelling place must be safeguarded for them, a dwelling place where we unite in us sky and earth, divinities and morals. A place where we do not simply invite to come visit us those who dwell far away, but where we discover as proper to us the near that lives in us and that remains foreign to us (Irigaray, 2002: 51).

Matrices of becoming

Life, as an ongoing project, involves carrying a constantly changing figure of the world within us. Were I to make a map from the cardinal points of my identities, it would resemble a chart of criss-crossing movements within continents and between hemispheres as well as dimensions. These are fast and strange times and we are moving in more dimensions than previously. Our habitat is technological and geographical; we live in a digital infrastructure as much as a physical one. Living in the culture of the contemporary technological habitat concepts of identity are no longer tethered to the earth but are in freefall as a multiplicity of becomings, hyper-realities and mixed-states.

One of the important aspects of going to see live performance is to be brought into proximity with embodied histories and to be inspired by the invention of new movement memories. In this way choreography is one way to incorporate and experiment with emergent realities. Within the technological theatre, the imaginary has a space to play and create that has not as yet always already been written upon by the globalising tendencies of mainstream art practices and the imperialising gestures of the past, including the dominance of the mainstream, including the malestream of a phallic imaginary. Performance events are generative in that they create spaces as action unfolds action, extending trajectories from layer to layer, point to point, contour to contour. As a woman dancing, I can operate a micro-politics of the self here by playing against and beyond phallomorphic constructions. As alternative body-forms and impossible anatomies emerge through my interactions with virtual spaces, we begin to inscribe a different history, a different morphology. The idea of matrixial becomings gains currency in this context.

Artist and Psychoanalyst Bracha L. Ettinger describes how the concept of the matrix shifts the womb from nature to culture, making it the basis for another kind of sense. In the matrixial borderspace, subjectivity is experienced as encounter, a place where *I* and *non-I* co-emerge and form a composite partial subjectivity. She sees this space as a place of passage, a 'Metramorphosis', a 'joint awakening of unthoughtful-knowledge on the borderline, as well as an inscription of the encounter in traces that open a space in and along the borderline itself' (2004: 77).

Through their experience of the womb, women can be said to have privileged access to understanding the matrixial borderspace as an event-encounter, as this corporeal dimension of their bodies provides an awareness of how 'outside meets inside'. Ettinger describes that although this privileged access to matrixial time and matrixial space could be a source of pleasure, in social and cultural terms, it is more often regarded as a source of fragility. It is important, however, to consider how a feminine construction of creative technology might access this 'archaic site of virtuality and potentiality' (2004: 77).

Screen technologies disseminate and proliferate images of the biological body to the extent that there is no longer a distinction between the inside and the outside of the body. The body's interior unfolded and exposed through machine vision, surveillance and medical technologies inform the image repertoire of the creative artist as she seeks to re-materialise the image. For what has been technologised is not the body itself but its image. Wresting the female body from its overdetermination as commodity, pornography and biology means becoming a 'midwife' for what Giorgio Agamben (2005) describes as 'this new body of humanity'.

The invention of a matrixial system that tells stories and creates different kinds of bodies might constitute a way to create an alternative identity that resists the violence of an imposed one. Working with emergent technologies which enable a play between the real and the virtual through an integrated circuit of bodies and technologies potentially fosters a negotiation and an encounter between dimensions. As a matrixial space, a place where something originates and develops, this space can allow for cosmographies of different kinds to co-exist and to operate in fertile interactions generating new forms and cartographies of the self. We can leave the ground without returning and this groundlessness is not necessarily a flight from who or what we are but a movement towards a different horizon and a home that 'lies ahead, in the unfolding of the story in the future, not behind waiting to be regained' (Warner, 1994: 88).

As a dancer, I perceive that embodied interfaces can potentially provide opportunities to experience a continuity of presences between the actual and the virtual, the real and the remote, the inside and the outside, the distant and the near, blending and rejoining spaces in a fertile, if at times untidy and confused, mix of 'through-otherness'. Such work can also potentially provide a platform through which to play through some of the contradictions of living cultural hybridity by creating the conditions for the simultaneous presence of different spaces, species and perceptions. But this is not without its risks for, as Jeanette Winterston states, 'when a fissure opens up in the self, half-known beasts climb out of it' (1997: 43).

Conclusion

We are and we are not our bodies. In dancing with creatures of code it is tempting to suggest that we are no longer confined by our bodies volume, weight, gravity and matter, that we are free to choose the extension of ourselves, to dance amongst the starfish of different skies, to play the puppetry of the virtual and to touch without also annihilating our rockpool phantasmagoria. But every moment of movement contains its own sightlines, just as every story I tell moves across another story. Dancing with our *virtual other* pushed the boundaries of our kinespheres, extending our movement in previously unthought of ways. The boundary of our bodies, their skin-sense and contour were mapped into the core of the virtual object as a black absence, in this way our outer limits became its inner limits, its animating core. Where does agency reside within this performance? The dancers were required to see and move from both the actual and the virtual point of view and to keep these positions in tension kinaesthetically and proprioceptively through feedback loops of interaction. Within these dynamic thresholds, the dancer is and is not present, just as the *virtual other* is and is not present. Embodied experience is reconfigured through interactions with virtual dimensions of space. From a feminist perspective, such experiences open up the potential for reimagining signification beyond the domination of an iconic femininity. So that rather than being 'cannibalised' by new technologies as Donna Harraway (1990) has stated, women can become agents of interaction inhabiting a plenitude of (dis)embodiments in which the palpably real and the ephemerally virtual co-emerge.[13] So that we give up taxonomies of difference and celebrate in patterns, rhythms, multiplicities, vibrations, pivots, joints, points of contact, crossings and energies of this new way of bodying forth.

Notes

1. Carol Brown, performance text, *The Changing Room*. Carol Brown Dances. Premiered 5 June 2004, Ludwig Forum: Aachen, Germany.
2. *Spawn* analyses the statistical characteristics of the silhouette's size and shape using the active shape models developed by Cootes and Taylor (Cootes *et al.*, 1995).
3. Seamus Heaney (2002: 366) writes that 'through-other' echoes the Irish-language expression, *tri na cheile*, meaning things mixed up among themselves. He uses this to describe the post-colonial condition of Irish poets in relation to Britain.
4. 'Other of the same' is a term used by Luce Irigaray to describe the logic and power of the phallologocentric mastercode which subsumes the minoritarian, the marginal and the other in its embrace. Braidotti (2002), with Irigaray, describes the differences proliferating in late postmodern or advanced capitalism as 'others' of the same in that the centre merely becomes fragmented. She suggests that rather than look at differences between cultures we explore differences *within* the same culture. Such a level of complexity would move beyond dualistic, oppositional thinking towards a new complexity which is transcultural and potentially transdimensional.
5. For a fuller description of these models and a guide to their use in experiential anatomy, see Olsen (1991).
6. This follows Luce Irigaray's intervention in phallocentric discourse. She defies the logic of normative definitions of identity which are unitary and which privilege the masculine. Her alternative figurations insist on a morphology which is not given but is made meaningful through practices of the self. See Irigaray (1985).
7. The *Changing Room* (2004) involved a collaborative team of Choreographer Carol Brown, Architect Mette Ramsgard Thomsem, programmers Chris Parker and Jesper Mortensen, Sound Design Jerome Soudan (mimetic), Lighting Design Michael Mannion, Production Management Gwen van Spijk, and the dancers were Catherine Bennett, Delphine Gaborit and Carol Brown. It premiered at the Ludwig Forum (Aachen, Germany) 5 June 2004 and was presented at Greenwich Dance Agency as part of Dance Umbrella, London, on 4–6 November 2004.
8. *Shelf Life* (1998), a collaboration with visual artist, Esther Rolinson, premiered at the De La Warr Pavilion, Bexhill-on-Sea, England, 3 December 1998.
9. See Wertheim, 1999: 171.
10. Deleuze and Guattari analyse the concept of *'becoming'* in *A Thousand Plateaus*, 1987: 232–309.

11. See Kathleen Woodward (2004) for a discussion of the history of emotional connections between 'lively machines' and bodies and the attributing of feeling to artificial lifeworlds.
12. Agamben, 2005: 49.
13. Ibid.

References

Agamben, Giorgio (2005) *The Coming Community.* Trans. Michael Hardt. Minneapolis: University of Minnesota.

Beckman, John (1998) 'Merge Invisible Layers', in John Beckman (ed.) *The Virtual Dimension.* New York: Princeton, pp. 1–17.

Braidotti, Rosi (2002) *Metamorphoses: Towards a Materialist Theory of Becoming.* Malden, MA: Polity.

Cootes, T.F., Cooper, D., Taylor, C.J. and Graham, J. (1995) 'Active Shape Models – Their Training and Application', *Computer Vision and Image Understanding,* Vol. 61(1): 38–59.

Deleuze, Gilles and Guattari, Félix (1987) *A Thousand Plateaus.* Trans. Brian Massumi. Minneapolis: University of Minnesota.

Ettinger, Bracha, L. (2004) 'Weaving a Woman Artist With-in the Matrixial Encounter-Event', in *Theory, Culture and Society,* Vol. 21(1): 69–93. London: Sage.

Harraway, Donna (1990) *Simians, Cyborgs and Women: The Reinvention of Nature.* London: Free Association Books.

Heaney, Seamus (2001) *Finders Keepers: Selected Prose 1971–2001.* London: Faber and Faber.

Irigaray, Luce (1985) *This Sex Which is Not One.* Trans. Catherine Porter. New York: Cornell University.

Irigaray, Luce (2002) *The Way of Love.* London and New York: Continuum.

Kwon, Miwon (2000) 'One Place After Another', in Erika Suderburg (ed.) *Space, Site, Intervention.* Minnesota: University of Minnesota, pp. 38–63.

Lefebvre, Henri (1991) *The Production of Space.* Trans. Donald Nicholson-Smith. Oxford: Blackwell.

Massumi, Brian (2002) *Parables of the Virtual.* Durham and London: Duke.

Olsen, Andrea (1991) *BodyStories: A Guide to Experiential Anatomy.* New York: Station Hill.

Smith, Daniel W. (1997) 'Introduction "A Life of Pure Immanence": Deleuze's "Critique et Clinique" ', Project in Gilles Deleuze (ed.) *Essays Critical and Clinical.* Minneapolis: University of Minnesota, pp. xi–liii.

Thomas, Sue (2004) *Travels in Virtuality.* New York: Raw Nerve Books.

Virilio, Paul (2002) *Crepuscular Dawn.* Los Angeles: Semiotext(e).

Warner, Marina (1994) *Managing Monsters: Six Myths of Our Time.* London: Vintage.

Wertheim, Margaret (1999) *The Pearly Gates of Cyberspace: A History of Space from Dante to the Internet.* London: Virago.

Winterston, Jeanette (1977) *Gut Symmetries.* London: Granta.

8
Kinaesthetic Traces Across Material Forms: Stretching the Screen's Stage

Gretchen Schiller

> The connection between the self and world grounded in the identity of the flow of the energy beyond and within the body enable videodance and filmdance – art forms that make us aware of the energy within the body even as they depict behaviour in the external world – to be a means for exploring the complex interrelations between motion and emotion, between action and affect, between an event's dynamic structure and its structure as a system of information.
>
> – Elder (1998: 298)

Introduction

Whether it is of bodies, light, sound, natural or simulated forces, movement has been, and continues to be, the subject of my choreographic research.[1] Today, this research results in artistic projects that transform qualitative movement vocabularies across material forms. These projects include videodance, performance and video projection installations. These works are choreographed specifically to elicit the general public's kinaesthetic awareness and to map new forms of movement perception.

In the installation *trajets*, which premiered in 2000, kinaesthetic relationships are choreographed, or *choreomediated*, among a specific architecture of screens and video images (camera, editing, post-production and real-time processing) for the general public.[2] This installation invites the public to stroll and wander in a kinaesthetically rich mediated environment with motorised floating screens: a body-screenography of sorts.

Within the context of this installation, both the general public and screens become dancers or players. Here, the visitor's dancing body is not overtly expressive but instead part of a movement dynamic that traverses trajectories in the installation. Together, the screens, images and the visitor dance and occupy the same artistic moving space. These relational interactions shift the attention away from the body and screen as separate objects, and instead emphasise the differential and relational play of movement tensions occurring between them. The participating visitor moves through the installation space, witnesses the behaviour of others and engages in an informal yet structured choreographic event. In other words, the visitor plays the role of dancer, choreographer and audience.

To achieve such an interactive dynamic, this installation integrates a variety of techniques including movement mapping, expressive screenography and principles of cinedance. Despite the contemporary nature of *trajets*, characteristics of these cited movement-based techniques reflect and echo scientific and artistic initiatives that evolved in France in the nineteenth century. This chapter explores these characteristics with special attention to the parallels that develop with the body screenographies of performer Loïe Fuller (1862–1928), the movement mapping by scientist Etienne Jules Marey (1830–1904), and the art form of cinedance emerging from both. Both researchers are relevant not only to the installation *trajets* but to my choreographic research in general. This discussion creates a historical context for contemporary and interdisciplinary movement research, which straddles the line between aesthetic and perceptual studies.

trajets

As visitors move through the installation *trajets*, the screens open and close around each individual. The physical presence and position of the visitor is captured[3] and, in turn, influences the ways in which the automated screens move in response to the visitor's movements. The rotational movement of each screen alters the space. The screens and visiting public together alter patterns in the installation space, which affect the screens' direction, velocity and rotation, and the visitors' pathways or movement trajectories. These interactions, which also include projected images collectively, constitute the choreography (organisation of the kinaesthetic tensions in time and space) of the installation. As the screens rotate, the room's fluid architecture shifts and the space dances . . .

Both American performing artist Loïe Fuller (1862–1928) and French scientist Etienne Jules Marey[4] (1830–1904) created techniques[5] that mediated and materialised bodily movement beyond its ephemeral ever-changing nature. Despite their disciplinary differences and

(a)   (b)

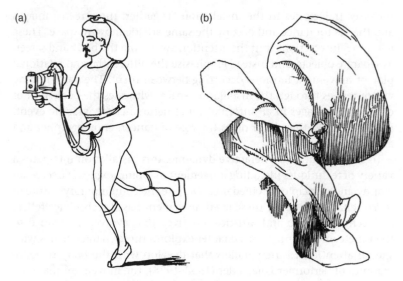

Figure 13   Original Sketches by Colin Lombard of images of Marey's experimental shoes and recording instrument ca. 1872 (a) and of Loïe Fuller dancing 'The Lily', ca. 1900 (b). These sketches were inspired from images of Marta Braun (1992: 27) and a photo by Isaiah West Taber (From Musée de l'Ecole de Nancy, 2002)

methodologies, both Fuller and Marey sought to expand the scientific and artistic scope of movement perception and transformation. Marey isolated the body into a controlled environment to capture quantitative shifts of movement over time with 'movement-mapping' techniques. Fuller, on the other hand, created an expressive 'body-screen', which transformed her body and its surrounding space into animalistic (serpentine) and elemental (fire) metaphors.

Independent of each other, they both contributed to *what constitutes movement and kinaesthetic knowledge*, one by framing, amplifying and converting movement into linear, sequential, two-dimensional representations, and the other by creating performances which transfigure the expressive materiality of bodily movement. They both can be seen as initiating cinedance (and videodance) traditions as we know them today.

Fuller's and Marey's seminal contributions to movement research are far beyond the scope of this chapter. What is of interest is how characteristics of movement perception in movement-based interactive art can be read historically and situated within a larger field of movement mapping and body screenographies, stemming from nineteenth-century research in art

(Fuller) and science (Marey). The cross-breeding of art and science in the domain of movement and dance is not only a product of the digital age.

Fuller and body screenographies

Loïe Fuller's performances premiered in Paris at the Folies Bergère in 1892. Fuller was raised in the United States, but her artistic career and professional identity were born in France. She extended the figurative body by transforming and choreographing movements with mechanical and electronic techniques and strategies. Fuller attached meters of white fabric to bamboo cane-shaped arm extensions to create a screen that enveloped her body. With this screen connected to her arms, she crafted and synchronised undulating flowing movement with coloured light projected onto the fabric.

Using these techniques, Fuller transformed her body into illusions of fire, animals and flowers. With a team of up to 27 electricians, Fuller choreographed movement relationships between the moving body-screen and real-time projection. Fuller's performance repertoire included titles such as La Serpentine: *The Dance of the serpentine*, The Violet: *La Violette* and The Butterfly: *Le Papillon* (1892). Here she transformed the figurative female body with bodily screenographies into metamorphic and kinaesthetic sculptures (Lista, 1994: 617–618) as the '. . . disparition du corps de la danseuse était en même temps nécessaire afin que le voile devienne lui-même l'expression du corps' [*The disappearance of the body of the dancer was necessary so that the screen became itself the expression of the body*] (Lista, 1994: 130, my translation).

Her innovative contributions went far beyond transfiguring the materiality of bodily movement with body-screens. Fuller conceptualised stages made with mirrors, which would reflect movement beyond the body and into the architecture of the theatre. Already, at the beginning of her career, she held four patents. These included: a new type of dress specially made for theatrical dance; a new form of stage design with optical illusions for theatrical dance (with lighting underneath the dancer); theatrical decorations made of a white wall garnished with stones; and a stage design with long mirrors upstage giving the space a polygonal appearance (my abbreviated translation from Lista, 1994: 618).

Marey and movement mapping

As Fuller was introducing herself to the French public, with her body screenographic techniques transfiguring the body, Marey was framing

and mapping the body by creating instruments to study the movement of animals, bodies and atmospheric elements. Busy in his laboratory the *Station Physiologique* in the Bois de Boulogne on the outskirts
of Paris, Marey developed instruments to frame, isolate, amplify, record
and analyse movement. The year that Fuller introduced performances
of a moving butterfly, Marey was studying insect wing movements.
In 1892 he published three articles: *Le vol des insectes étudiés par la photo-
chronographie, Le mouvement des êtres microscopiques analysé par la chronophotographie* and *Le mouvement du coeur, étudié par la chronophotographie.*
The following year he published a paper on the movements of the
swimming stingray (Braun, 1992: 434). Independent of each other, yet
belonging to the same cultural and kinaesthetic climate, Marey and
Fuller were materialising new forms of movement perception converging
the movements of animals, the environment and humans. They were
mapping new movement perceptions and aesthetics.

According to Marey, humans were ill-equipped to perceive movement.
He felt that recording instruments should be created and considered
as 'new senses of astonishing precision' (Braun: 40). Contrary to many
of his colleagues, Marey defended and promoted movement perception
as a viable research topic in the French scientific academy. Building
upon Carl Ludwig's kymograph (1847), Marey began by building the
sphygmograph (1860): 'The sphygmograph transformed the subjective
character of pulse feeling into an objective, visual, graphic representation that was a permanent record of the transient event...' (Braun,
1992: 18).[6]

Marey transformed movement into a material 'trajectoire spatiale' or
'enveloppe temporelle' of continuous points in space (Didi-Huberman,
2004: 245) and continued to build instruments to study the movement of animals, people and the environment. He felt it necessary to
supplement the human inability to perceive the ever-changing nature
of movement. Marey wanted to isolate movement and detach it from its
spatial and temporal condition in order to better visualise and analyse its
quantitative nature by optically marking the human body, connecting
air pumps to horse's hooves, and analysing trajectories of air current.
His real-time instruments mapped movement onto two-dimensional
graphs and pictures. Creating the *screened body*, his instruments became
mutually constitutive processes of movement perception. Or, as visual
and cultural critic Cartwright states,

As the horse's body motored the inscription device, so the kymograph
inscription reconfigured the conception of the living body from

within, rendering it an ordered living system – a system best represented by graphical, temporal forms like the calibrated kymographic line or the incremental cinematic image for example. (Cartwright, 1995: 24–26)

trajets

Like Marey, the installation *trajets* uses mutually constitutive processes of bodily movement techniques to map feedback encouraging the visitor to feel movement vs simply looking at movement.[7] Mapping in *trajets*, however, occurs at different levels of interaction and materiality, including real-time feedback and traces of pathways.

First, the integration of real-time feedback and processing with computer techniques in *trajets* significantly reduces the gap between action and representation. Inanimate objects (the screens) become animated and generate choreographic and participatory movement rapports between the visiting public and a mediated environment. This practice focuses on dynamic action-response, feedback and kinaesthetic inscription facilitated through the use of digital, optical and mechanical techniques. Visitors improvise in a structured, choreographed space with simple or common movements (weight shift, stopping, walking) in response to the technically mediated conditions of the installation.

The screen is not only a projection surface, but also a dynamic participant in the screenography, with video and force-based reactions that move in relationship to the visitor in real-time. The screens in *trajets* rotate in response to the visitor's physical displacement and position in relationship to each screen. The interactive experience is articulated through motorised suspended screens moving in response to the visitor's path: the screens spin and twitch in response to the visitor's body. The screens in *trajets* do not separate the subject of the visitor's movement experience from its *representation*, but instead seek to develop a participatory dynamic which continuously maps and *renders present* movement perception between the participant and the given feedback experience.The visitor experiences continuous movement feedback as well as interactive surprises and unfamiliar movement reflexes. In other words, the screens' movements introduce the visitor to alternative pathways to follow, and movements with which to play or dance.

Additionally, visitors' pathways are recorded in real time but then projected in alternative spaces and in different material forms. In the 2000–2002 tour of *trajets*, a few forms of remote visual mapping techniques were explored. In one visual mapping, each visitor was represented by a different-coloured moving dot. This dynamic visual representation of movement was projected outside the installation and it served as a referencing map both to recall (for those who have visited the installations) and create a prologue to (for those still waiting to enter the installation) the idea of movement trajectories. This mapping built upon the idea of extending the visitors' movements to different locations as trajectories crossing various spaces. As such, each visitor activated and left traces of movement in space. This mapping of pathways suggests the ways in which we leave traces in physical space conjugating the physical space we inhabit.

Like Fuller's choreographies or screenographies, the screens in *trajets* dance. Unlike Fuller's prosthetic screen or body-screen, the screens in *trajets* themselves are not attached to a body, but instead are electronically and digitally choreographed with stepper motors. Regardless of this difference, however, the visitor experiences a common kinaesthetic pull or sensation of movement. The movement of the screens ripples around the bodies of the visitors like waves propagating around a body in water, like Fuller's undulating screen.

When watching Arnaud Esterez' film *La Loïe Fuller*, I felt a common physical link between Fuller's undulating screen and the swerving screens in *trajets* (see Esterez, 1999). A common kinaesthetic climate of undulation, suspension, curving, flowing, swooping. Like Fuller's screenographies, the projected images transform the qualitative nature of the space. Like Fuller's screens, the visiting public and screens in *trajets* carve space. Like Fuller's body, the visitor's body and the physical space together appear as metamorphic and dimensional trajectories in space.

In a similar vein as Fuller, the projections of the video images onto the moving screens in this installation transform the physical space. The physical space takes on an ensemble of metamorphic states. The visitor's body in *trajets* has the opportunity to experience its own inner-felt movements' *kinaesthesia*, a sense of empathy for other participant's movements and/or physical connection to the images.

Cinedance

The body–space screenographies in *trajets* are not only reflective of nineteenth-century gestures in science and art, but also strongly influenced by cinedance and videodance practice and theory extended from these nineteenth-century inventions. Cinedance traditionally has a seated and physically passive audience. Cinedance is an art form that extends both what is cinema and what is dance.[8] This art form extends the body-medium of the audience through movement empathy and haptic perception with cinematic techniques of shifts in point of view, referential framing, décor, montage, compositing and so on. The audience's empathetic identification with moving images is a sort of haptic or kinaesthetic inscription. A film inscribes sensations into the seemingly passive seated audience member. The public is transported physically to alternative physical spaces without travelling or dancing physically themselves.

Examples of this art form include *Ballet Méchanique* (1924) by Fernando Leger, in which 300 images of people and objects are edited together at a very high speed, Leni Riefenstahl's *Diving Sequence* from the (1936) Berlin Olympics, and *New York New York* by Francis Thompson (1958) where the kaleidoscope prisms, wide angle lens and mirror create liquid bodies moving through a fluid movement dance of architecture of the city of New York.

These movement-based art forms create dynamic and corporeal links between the art form and the public's physical reception and perception. Here the technical processes co-construct one's kinaesthetic repertoire of movement sensation and experience. Visual and cultural critic Marks describes this physical internal reception of the public as haptic cinema – a sort of visceral chiasmus of perception and moving images.

> Haptic cinema does not invite identification with a figure – a sensory-motor reaction – so much as it encourages a bodily relationship between the viewer and the image. Consequently, as in the mimetic relationship, it is not proper to speak of the object of a haptic look as to speak of a *dynamic subjectivity* between looker and image. Because haptic visuality tends less to isolate and focus upon objects than simply to be co-present with them ... I am suggesting that a sensuous response may be elicited without abstraction, through the mimetic relationship between the perceiver and a sensuous object. (Marks, 2000: 164, emphasis added)

trajets calls upon the visitor's 'dynamic subjectivity'. In the installation, the projected video images are at times figurative (one recognises the body – in a mimetic manner) and at other times more abstract, showing traces of bodily movement. *trajets*, like Fuller's body-screen projections, questions the visitor's visceral notion of *what is body – what is space – what is image and what is screen* by transcending the figurative body into a sensual and moving dynamic.

One of the members of the public participating in the exhibition of *trajets* at the Institute of Contemporary Art (ICA), London 2001, spent approximately one hour in the installation dancing and improvising with the screens. At times he moved forwards and at others backwards. He changed his body-shape in relationship to the screens' movements and played rhythmically with each screen's movement quality. Before he left, I went up to him and complemented his movement improvisation. He told me that he was not a dancer (and here I paraphrase) and that the installation gave him the opportunity to explore spaces, which opened, closed, pulled, caught, tricked and surprised him.

The technical-kinaesthetic-inscription characteristics of this research bring forth new movement rapports and relational processes through interaction of felt movement. They do not replace the body, rather they intend to stimulate and anchor the participants in their *kinaesthetically aware body*. At the same time, this installation research introduces new movement interactions within a movement field of moving objects, other participants and the given choreomediated environment.

Current research

This research of interactive, choreographed screenographies and movement mapping is currently in the process of exploring new movement tracking and screenographic techniques. This research involves more physical interactions and surprises between the screens, the images and the visitor's potential movement exchanges within the space. The choreographic interaction still depends on the physical participation of the general non-specialised public, but unlike *trajets* 2000 it no longer uses computer vision to track the movement of visitors in the installation. In order to capture more people simultaneously and with more accuracy, the team pursued the design of a pressure-sensing floor.[9] The technical and artistic goals of this project have become more specific and complex. These include synchronised multiple projections, a choreographic system design and variations of behavioural interactivity.

The research is developing translations of dynamic and dimensional mapping. Mapping is not limited to a continuous linear series of dots or curves in space on a two-dimensional image (Marey). Both individuals and groupings of people influence the nature of the installation. The intent is that the visitor is not only aware of his or her body, but of a movement field of experiences or lived trajectories through space.

The screenographic research involves investigating ways in which each screen is its own body-screen, part of a community of body-screens mimetic of the participant's body. A screen can move on its own or it can belong to a set of movement behaviours with other screens. Individual and groups of screens receive and reflect video projections. At times the screens are double-sided, each side receiving a different video projection.

Conclusion

This choreographic research maps both familiar and new movement perception and stretches the screen's stage. Movement mapping, body screenographies and cinedance principles contribute to the choreographic research. Together they create a dynamic movement field of interactions or a 'kinesfield' within which the general public has the opportunity to become aware of movement within and around their bodies.[10]

This artistic research also indirectly reflects the microcosm of a larger, technically mediated socio-kinaesthetic condition we experience in daily routines. Vibrations from the car, acceleration in a plane, deceleration in a train or a descent in an elevator are all kinaesthetically felt, technically mediated experiences. These technical systems intersect continuously with our sense of movement inside, around and far away from our bodies. Whether apprehended passively or actively, these experiences contribute to the range of one's movement repertoire and kinaesthetic condition. Every technically mediated action we undertake, or with which we are engaged, is embodied. We may not, however, be conscious of this perpetual technical-kinaesthetic inscription which takes place in the movement field we live in. *trajets* draws attention to these embodied technically mediated movement transactions.

Technical systems are not separate from movement perception, but entangled in the manner in which we come to experience, understand and perceive movement: they constitute our perceptual kinaesthetic systems. If we accept this entanglement between human-created techniques and movement as a dynamic structural and relational event, then we replace discussions of the *body and space* or *body and machine* with the fluid surprises of relational dynamics.

Notes

1. For instance, works such as: *Face à face*, a videodance performance (1995), *Camarà*, a videodance documentary (1996), *Suspended Ties*, a videodance (1997), and *Shifting Ground*, an interactive installation (1999). See http://www.mo-vi-da.org.
2. *trajets* is co-directed with Susan Kozel (2000). An ensemble of experts – Robb Lovell (computer programmer and performer), Jonny Clark (composer), Pablo Mochcovsky (engineer), Shaun Roth (architect) and Scott Wilson (computer programmer and system designer) are members of the *trajets* team.

3. ̄ *trajets* 2000 used computer vision as a movement-sensing technique.
4. Edward James Muybridge, a contemporary of Marey, is not discussed in this chapter; however, his work and collaborations with Marey contributed to moving image techniques.
5. I choose to use the term 'technique' instead of the term 'technology'. A technology is the discourse relating to a technique – often the term 'technology' is used when we really mean 'technique'. See Roberto Barbanti (2004) *Visions techniciennes: de l'ultramédialité dans l'art*. Nimes: Théétète.
6. Carl Ludwig (1816–1895) introduced this first graphing device to physiology (Braun, 1992: 18).
7. There are correspondences between motion capture 'mocap' techniques and Marey's optical motion tracking systems which are unfortunately beyond the scope of this chapter.
8. In the 1960s, choreographer, ethnographer and filmmaker Allegra Fuller Snyder distinguished three categories of dance and film. These included the *dance filmic document* (i.e. single camera point of view of the staged dance performance), *the dance translation* (i.e. the narrative of the dance is respected but adapted for the camera with close-ups, three camera shoots, etc.) and *cinedance* (i.e. the creation of a new art form which transcends the biological possibilities of the living, gravitational body and which introduces alternative forms of embodiment). For more, see Snyder (1965).
9. Pablo Mochcovsky was the head hardware designer and engineer for the new pressure sensitive floor being used in the new version of *trajets* which has been supported by incult http//www.incult.es.
10. The kinesfield describes this interactive process of one's embodied state as relational to its environment through temporal and spatial phenomenological (subjectively felt) dynamic transactions. This chapter includes sections from my doctoral thesis. For more, see *The Kinesfield: A Study of Movement-based Interactive and Choreographic Art* (Schiller, 2003).

References

Barbanti, Roberto (2004) *Visions techniciennes: de l'ultramédialité dans l'art*. Nimes: Théétète.

Braun, Marta (1992) *Picturing Time: The Work of Etienne-Jules Marey*. Chicago: The University of Chicago Press.

Cartwright, Lisa (1995) *Screening the Body: Tracing Medicine's Visual Culture*. Minneapolis: University of Minnesota Press.

Didi-Huberman, Georges (2004) *Mouvements de l'air. Etienne-Jules Marey, Photographe des Fluides*. Paris: Editions Gallimard.

Elder, R. Bruce (1998) *The Body of Vision: Representations of the Body in Recent Film and Poetry*. Waterloo: Wilfrid Laurier University Press.

Esterez, Arnaud (1999) *La Loïe Fuller*. 9′ Colour VHS. Angers: A Capella productions, France 2 TV 10 CNC, Dancer: Claire Dupont.

Lista, Giovanni (1994) *Loïe Fuller Danseuse de la Belle Epoque*. Paris: Stock, Editions d'art Somogy.

Marks, Laura (2000) *The Skin of the Flesh. Intercultural Cinema, Embodiment and the Senses*. Durham: Duke University Press.

Musée de l'Ecole de Nancy (2002) *Loïe Fuller Danseuse de l'art nouveau*. Paris: Editions de la Réunion des musées nationaux.

Schiller, Gretchen (2003) *The Kinesfield: A Study of Movement-based Interactive and Choreographic Art*. Unpublished doctoral thesis. UK: University of Plymouth.

Snyder, Allegra Fuller (1965) 'Three Kinds of Dance Film', *Dance Magazine*, 39: 34–39.

Websites

Trajets Tour (2002) http://www.trajets.net.

Mo-vi-da http://www.mo-vi-da.org

9

Sensuous Geographies and Other Installations: Interfacing the Body and Technology

Sarah Rubidge

I am approaching this chapter as a practitioner-scholar whose primary practice lies in the field of immersive installations. All have as a central feature the interface between the body and technology. In this chapter I will explore the nature of the relationship which occurs between body and technology when the participant engages experientially with immersive installations. I will also ponder on the way in which Gilles Deleuze's and Henri Bergson's thought finds voice in my work, and make explicit the way in which recent findings in neuroscience have provided an unexpected scientific underpinning for the approaches I take with respect to my installations.[1]

Talking about immersive interactive installations is not easy. They are primarily multi-sensory *experiential* environments, designed to be inhabited rather than viewed as an artefact or event.[2] Because the viewers are responsible for generating and/or processing imagery in real time, each material manifestation of these installations is unique, a never-to-be-repeated event. Like multiscreen immersive video installations the 'art' of such installations does not lie solely in their material presence, but also in the dynamic architecture generated in the 'space-in-between' of the material architecture and the images, that is, in the spatial and temporal contrapuntal rhythms which are composed through the ever-changing, ephemeral motion of the images on the screen (Morse, 1997).

In the immersive interactive installation, the participating 'viewers' themselves become active elements in the installation environment, responsible both for the initiation of individual image elements and the modulation or inflection of the intensities from which they are composed. This gives rise to an interchange both within and between the individual images as the choreographed progression of each image implicitly interacts to create a collective spatiotemporal event. In the immersive installations these intensities are not merely seen, not merely heard but experienced through what Paul Rodoway (Rodoway, 1994) calls the intimate senses (the haptic, the kinaesthetic, the visceral, the proprioceptive). It is as much through these as the 'distant senses' (hearing and sight) that the works reveal themselves to the viewer/participant. As a result, the physical manifestation of the install-ation, that which is observed or heard, is not their *raison d'etre*. Rather it is the sense of being, or becoming, that the participants experience through their deeper physiological responses as they become absorbed into the installation. In short, immersive installations are not simply 'pictures' or images to be viewed from the outside, which artefactu-alises the installations, but are essentially ephemeral and experiential events.

In spite of the difficulties of representing such an event, in this chapter I will be offering images of two immersive installations. These images do not, indeed cannot, address the experiential factors which lie at its heart. Despite Henri Bergson's observation that 'well-defined outlines [of the word] . . . overwhelm or at least cover up the delicate and fugitive impressions of our individual consciousness' (Bergson, 1910: 132) and thus offer an inadequate understanding of our experience, I turn to words in this chapter. Here I will try to capture something of what lies at the heart of the relationship between the body and technology that immersive installations embody, to reveal the sense of being, of becoming, that permeates them.

As an artist my work for the last 12 years has been predicated on the belief that an interface between complex responsive systems devised with new technologies and the more subtle operations of the human body can illuminate for both the artist and the public, our more subtle ways of 'being in the world'. Inevitably this has led me to engage with writings from both philosophy and science which have addressed that which underlies the sensate body. Such writings acknowledge that the information which comes in through pre-reflective bodily responses constitutes a form of consciousness and a mode of knowing which are as valuable as the kinds of knowledge which are generated by a reflective consciousness mediated by words or other symbols.

Neuroscientists such as Damasio, Claxton and Edelman, for example, argue that consciousness has several levels, and that those which lie beneath the level of reflective ('higher' or 'extended') conscious-ness are as crucial to understanding as reflective consciousness. They suggest that the underlying modes of consciousness, which Claxton calls 'the undermind', Damasio 'core consciousness', and Edelman 'primary consciousness' not only provide us with crucial information about the world, but also are the source of intuitive insights, which we generally consider to be outside of the remit of what we understand to be 'consciousness' (Claxton, 1997; Damasio, 2001; Edelman and Tononi, 2001). These non-verbal modes of knowing are the business of the autonomous and semi-autonomous physiological systems which are inextricably implicated in embodiment.

Since 1994 I have been exploring the interface between the body and new technologies. Initially I was interested merely in introducing the sensuous into the digital domain, which was at that time dominated by computer graphics with a decidedly non-sensuous quality.[3] Later I became increasingly interested in the body as interface to technology. In *Passing Phases* (Rubidge *et al.*, 1995–99), my first interactive installation, the interface lay between audience members and a multiplicity of visual images shown on monitors mounted on a circle of white plinths. The displays on the monitors comprised a progression of moving images, selected at random from a bank of video close-ups of human hands, feet, lips, tongues.[4] They stroked, touched, grasped each other, now slowly, sensuously, sensitively, now vigorously, nervously, aggressively. Moving variously in unison and in canon, the images were always changing, revealing variations of actions, and of the emotions associated with them. I was operating on the underlying principle that the very act of viewing the intimate actions of other human beings at close quarters could set up a physical, or more accurately a physiological, resonance in the viewers.

It seems that here I was anticipating the work of a group of neuroscientists in Italy who, in 1996, discovered a group of neurons in the frontal cortex called 'mirror neurons' (Gallese *et al.*, 1996). They found that the same patterning of neuronal activity is activated in these neurons whether the action is being viewed or enacted by the subject. A similar principle has been accepted intuitively in the field of dance for several decades. It has long been held that our understanding of a dance performance emerges from something deeper than cognitive understanding. It is implicit in the claim made by Martha Graham that 'movement never lies' (Graham, 1991: 4) and in John Martin's notion of *metakinesis* (Martin, 1939). It is also implicit in the principles upon which somatic movement practices such as Body Mind Centring and Release Techniques were based. The installations I make are a continuation of these perspectives. Like many dance practitioners of my generation, the 'bodywork' developed by the artists who made up New York's 1960s Judson Church Dance Theatre and Britain's 'New Dance' movement of the 1970s and 1980s has permeated my artistic sensibilities. In installations such as *Passing Phases* and *Time & Tide* (Rubidge and Rees, 2001) these notions are implicitly invoked.

In pursuance of this intuition *Time & Tide* used visual imagery based on human motion (two women rolling back and forth in the shallows of a rural harbour).[5] This imagery, however, was abstracted to such a degree that the corporeality of the body in movement was diminished until almost nothing but the flow of motion through the body remained. The original imagery was processed by saturating raw video material with resonant blues and by merging, stretching, reducing images to mere traces of the texture of the original until all that remained when it was projected onto the installation structures was the flow of colour and motion as the diffuse images of the bodies rolled back and forth, this way.[6] The composite imagery was then distributed across a space broken up by multifaceted structures, fragmenting any traces of representation which remained.

In *Time & Tide*, as a 'viewer', you are immersed in the 'material' manifestations of the technological environment;

immersed in a room full of irridescent blue light, a drifting
soundscape, the smell of the
seashore . . . as your eyes slowly adjust, the blue light
flows over the surfaces of a cluster of ragged pyramids
which seem both to absorb the
flickering light and deflect it onto walls, ceiling, sand . . .
unpredictably a hazy image of a woman clad in a long clinging
garment momentarily
appears on the face of one of the pyramids . . .
and then is gone as if she had never been . . . the space feels
'calming, serene . . . eerie and sensual'

. . . you walk further into this mysterious environment, move around
the cluster of
pyramids, always one step behind these tantalisingly brief appearances
of human forms
which appear momentarily in
different parts of the space, barely leaving time for them
to register in your consciousness . . . everything is
'near and far, total yet partial, delicate, fragile'

... the light plays across the sand, momentarily revealing a sculpted
arm, leg, face, an
    encrusted bottle, a cuttlefish as the shifting light
pattern flows back and forth in a continuous, yet discontinuous,
    motion... the luminosity and texture of the blues change
    intermittently... now light, now dark, now glowing,
        now a thunderous, deep, angry green...
    the smell begins to fade... the sound lulls you... it is
    'primordial, mysterious, beautiful, mesmerising...
        peaceful and meditative'
You leave the space and suddenly are in the light, where everything
    has boundaries,
            shape, form, definition...
the workings of your mind begin to replace the workings of sensation.[7]

The intention in *Time & Tide* was that content of the visual imagery
would lie on the threshold of perception, inaccessible to extended
consciousness, accessible only to the 'undermind', to the psychophysical state which obtains immediately before an image is 'grasped' and
accrues a cognitive presence (Pelli, 1997).

More immersive than *Passing Phases*, perhaps because the majority
of the imagery lay on the threshold of perception, *Time & Tide* gave
substance to the philosophical ideas of Bergson and Deleuze which were
beginning to permeate my thinking at that time, and were increasingly
affecting the texture of my work. Indeed, it was during the making
of *Time & Tide* that it became apparent to me that my artistic work
was increasingly predicated on the desire to set up the conditions for
becoming. As both artwork and event the installations never reach resolution, are always in process. Composed of a collection of intersecting
complex systems they increasingly present visual and/or sonic imagery
whose content lies on the threshold of perception. The imagery is
constantly changing shape, quality or texture in unpredictable patterns,
looped sound and/or visual images modulate in a phased rhythm, setting
up differential relationships which are continuously shifting, and thus
generating ever new textures, ever new dynamic forms. All this leads to
a sense of motion, of a flowing hither and thither, focused but without
precise definition and never coming to rest, never reaching fulfilment.
This is a becoming which is not concerned with becoming anything,
or with reaching a particular state of being, but is simply 'absorbed in
occupying its field of potential' (Massumi, 2002: 6).

The resonances of Bergson's and Deleuze's thought within my work cannot be ignored. Two aspects of their thinking are deeply implicated in my artistic processes, and the nature of the work I produce. These are the notion that entities constitute 'qualitative multiplicities' and the notions of 'affect' and 'sensation'. The former is an analogy for the structuring processes and the 'work-systems' which underlie complex interactive installation works, the latter provides a potential means of beginning to understand the nature of the experiences the works are intended to facilitate.[8] These ideas have a particularly deep relevance to the work of artists involved in generating multi-user interactive installations. An individual participant's responses are not only affective responses, but also necessarily part of the more complex interweaving network of responses generated by an interacting group of individuals which is, in Deleuze's terms, a qualitative multiplicity.

Both Bergson and Deleuze talk about the nature of 'reality' as being more akin to a qualitative than a quantitative multiplicity. Whilst the latter is static, homogenous and spatial, the former is heterogeneous, temporal and nomadic. Consequently, the elements of a quantitative multiplicity stabilise into a fixed entity with a definable shape (e.g. object, artefact), whereas qualitative multiplicities (e.g. physiological systems) are dynamic temporal entities, constantly in flux, always reforming as new relations are initiated within them.

The notion of the qualitative multiplicity is intrinsic to multi-user, multi-modal, interactive installations. As in a qualitative multiplicity the movement of the elements of such an installation is continuous and relational, the elements interpenetrate one another, becoming qualitatively other through the particularity of their ever-shifting relational dynamic. Like a qualitative multiplicity they cannot be subject only to material analysis, as can stable entities, for their very qualities are always in flux. Made up of an array of fluid elements, which are constantly mobile, constantly merging into and out of new yet temporary formations, with an attendant qualitative modulation as elements interpenetrate and resonate with each other, there is no 'form' *per se*, only a system which is in constant movement. Any 'territorialisation' is fleeting, as other factors are continually brought to bear on the dynamic systems which might have been temporarily stabilised, effecting a destabilisation of what appears to be a 'form' and a recommencement of the complex network of movement that characterises it.

Interactive installations are constituted by a complex internal dynamic computer system. Whilst their features and the output they generate can be itemised (given a place in a 'material' description of the installation), this is not the 'work-system' itself.[9] That lies equally in the interpenetrating *behavioural potentialities* of the different elements, strands and planes of the system, which whilst programmed into them are not predetermined, most being built on the principles of Boolean and/or fuzzy logic.[10] The internal behaviours of given strands (say sound or visual strands) set in motion the particularities of both the composite behaviour of the complex system within which they operate at any given moment (its 'state', if you like) and the behaviour of the output of the strand itself. Thus a behaviour of an element of one strand of the installation system will initiate behaviours in another strand or strands of the system, which in turn will initiate a modification in another strand of the system, which might have a return impact on the first strand, and a new impact on others. Thus, whilst a behavioural 'decision' in one strand of the system may be simple, the impact of that decision on other strands is infinitely complex, and cannot be predicted precisely, because the specific composition and dynamic relations of the latter will be different from one moment to the next. These ever-changing dynamic relations between elements within the system constantly change the weighting of its dynamic structures both locally and globally, and thus the balance and features of the sonic and visual output of the installation.

Ideally, complex multi-user interactive installations will have been built in such a way that for the participant player(s) there is a certain legibility to the system's local responses to their behaviours. This, notwithstanding the responses of the system itself, goes far beyond the local, as the complexities of the impact change in one strand of the system and build on other strands. The shifting interactions between the players and the system become the framework for a collective 'event'.

Interestingly, whereas the behaviour of a (non-AI) computer-based system theoretically has limits, the behaviour of one of the most important strands within the interactive installation system, that is the behaviours of those who engage with it, does not have such limits. The participants in a multi-user interactive installation are the 'wild cards' in the system, for their understanding of the system is gleaned from a variety of prior experiences of both life and art, and from the deeper levels of their physiology. This influences their responses to the installation's behaviour, and frequently disrupts the expectations of the designer(s) of the system by doing something the latter had not anticipated, and thus setting the system off into unexpected directions. This is both the joy and the frustration of creating multi-user interactive installations.

Sensuous Geographies (Rubidge and MacDonald, 2003) is one such installation. It comprises a large installation space hung with translucent banners, upon which abstract digital figures move. In the centre of this space is a 4 m × 4 m circular floorcloth, which marks the boundaries of the (inter)active space. Above this space is a camera which captures the motion of those who enter the space using colour-tracking software. This allows a sound strand first to be initiated and then modulated by each individual player. The trajectories of the players' motion is the main parameter for the modulation and spatialisation of the individual sound strands in real time.[11] Visitors to the installation can choose to enter the space wearing a full-length silk robe in red, yellow, green or blue through which they are individuated by the system. This allows them to interact individually with the installation.[12]

The number of permutations of the interrelationship between the sound strands generated by four players is immense. For this reason the *Sensuous Geographies* work-system has a number of levels of complexity so that the players can engage with the installation at different levels of expertise. The simplest level (for the novice player) entails the spatialisation and sonic modulation of single sound strands through single-player behaviour. The more complex (expert) levels entail modulation of the sound strands through 'group' interactivity, for example through variations in the proximities between two and four players. These latter levels of interactivity are sometimes impossible to 'read' consciously. However, when a user reaches this level they have become skilled at 'reading' the environment and noticing differences in the responses. Also, they tend to respond with the intuitive understanding of the installation environment which is detected by the undermind rather than by a conscious understanding of the results of their actions.

... you tentatively enter the space, stepping over the threshold
between
inner and outer, your eyesight masked by a veil... It is
'like entering a magical space, a Prospero's Island and existing wholly
within it'
You find yourself navigating through the space using your hearing,
understanding the volume and shape of the space through the
movement of the sound. But you don't 'understand'...
you leave the space, stand on the edge.... you raise your veil and
watch
others moving in the space, 'dancing' gently or vigorously as they
try to change the quality of their sound strand...
alone, or as if in a blind quadrille.

You let the veil fall over your eyes and re-enter the space . . . you relax,
listen to your body, let the kinaesthetic sensations you find yourself
experiencing through the pitch, the volume, the texture
of the sound drive your movement . . .
'feet on soft floor clinging to the noise that is mine' . . .

you find yourself unexpectedly rising onto your tiptoes, leaning
forwards, riding the high-pitched sound . . . experiencing
'an impulse to move, a reason to stay – at the threshold of perception'.
'It is intriguing, elemental, absorbing' . . .
you listen to the shifting spatial features and textures of the sound
environment and 'open an ear to [your] body' . . .
you 'suddenly get ears and feet again . . .
new thoughts about consciousness'

the bright underlying sound you are riding is replaced by a deep
grumbling sound . . . your find your movement subtly changing, it is
slower deeper . . . You are 'in a different state (of consciousness?)' . . .
The 'undermind' has taken over . . .[13]

The responses noted above, which are both conceptual and physiological, were those of individual players. The collective response of a group
of participants to the sonic texture of the environment, however, plays
an equally important role. The collective response operates at several
levels, becoming increasingly deepened as the experience of the players
with the installation increases. Initially players become aware that they
can hear not only their own sound and its pathway, but also the pathway
of other sounds. A sound moving towards a player indicates that another
participant is approaching, and vice versa. They begin to play with this,
deliberately 'following' other participants, moving towards and away
from them, sensing their shifting proximities. Later they might realise
that the proximities themselves are the source of some modulations of
the sounds. Unwittingly as they respond to the very local conditions to
which they have access they create an emergent group 'choreography'.[14]
Like the installation system this collective behaviour is itself a dynamic
system, characterised by spatial dynamics, which are constantly in flux,
constantly generating compositional modulations in both the installation environment and the movement event which is taking place.

As can be imagined, the complexity of the interrelationships between
the three main systems which are inherent in an activated *Sensuous
Geographies* – the computer system, the group-behaviour system and
the individual player's physiological system – is such that it is almost

impossible for participants to grasp the logic of the computer system conceptually whilst fully engaging with it. This becomes for some players a matter of concern. The system, however, is set up to allow players to respond in a variety of ways, either through deliberate, cognitive decisions or through non-cognitive responses. Many participants moved happily back and forth between cognitive decisions and non-cognitive responses. However, those who have found themselves able to relinquish the need to understand conceptually find that their responses to the installation are driven by their physiological, rather than conceptual, responses, and are thus frequently in the hands of the 'undermind' and thus beyond their conscious control. And it is here that the second aspect of Bergson and Deleuze's philosophies find their place in my work.

For Deleuze the notion of sensation reaches beyond the overt physical sensations that we feel in our bodies to encompass the notion of sensation as be(com)ing. It has already been noted that the works I create are intended to set up the conditions for becoming. In this I had unwittingly taken up Deleuze's suggestion that the work of art is the 'being of sensation' (Deleuze and Guattari, 1994: 165). Deleuzian 'sensation' is constituted by the mobile forces and intensities which comprise, in part, the 'work' of an artwork.[15] Any representative power is subsumed under these intensities. The mobile forces which permeate a work of art which is redolent with a Deleuzian 'sensation' continuously 'blend into one another in subtle transitions, decompose, hardly glimpsed' (Deleuze and Guattari, 1994: 186), ever on the threshold of surface consciousness as we experience the work, but never materialised as a fully formed, resolved expression. Beyond representation, beyond expression, these forces and intensities are felt as states by the visitors as they simultaneously generate and experience the immersive installations I have discussed in this chapter.

Intimately tied up with this notion of 'sensation' are what Deleuze refers to as 'percept' and 'affect'. Deleuze distinguishes between the modes of understanding he identifies as belonging to the domains of the 'concept', the 'percept' and the 'affect' (Deleuze and Guattari, 1994). The 'percept' and 'affect' are implicated in the modes of knowing and understanding which are the business of the arts, particularly those works which embrace ambiguity and a degree of indistinctness in their imagery. Whilst the 'concept' is 'an act of thought' (Deleuze and Guattari, 1994: 21), and the domain of philosophy, the percept goes beyond thought, beyond even the domain of perception. Similarly the affect goes beyond the domain of feeling. In the Deleuzian percept the perceiver discerns the relations between a complex of intensities, forces, experiences which are brought forth by a perception but go beyond mere perception (and here a connection with Morse's notion of the site of the 'art' of the installation can be identified). This percept is 'a perception in becoming', one which transcends the discernment of determinate milieux replacing it with a perception of the indiscernible 'nature' of the perceived.

The affect like the percept goes beyond the condition of being 'affected' by a phenomenon, beyond emotional or even sensual response.[16] Affect is felt, but not as a specific sensation, nor as a specific feeling. Rather it makes itself known in a zone of indiscernibility as a 'state of suspense' (Massumi, 2002). Massumi suggests that it is an ephemeral sense of 'passing into', of becoming intimately attuned to the intensities of the experienced phenomenon, of bypassing the need to understand it in terms of representation or expression. It is this experience I aspire to in my installations.

That Massumi argues that affect is analogous to the 'intensity' which characterises a phenomenon, a work of art, a natural force is worthy of note. It is, he argues,

> [E]mbodied in purely autonomic reactions most directly manifested in the skin – at the surface of the body, at its interface with things . . . outside expectation and adaptation [and thus intention] . . . spreading over the generalized body surface, like a lateral backwash . . . travelling the vertical path between head and heart. (Massumi, 2002: 25)

Although this is undeniably in the right domain, the autonomic reactions are not always most directly manifested in the skin. They are manifested in every system which make up the human organism, from the hormonal, the circulatory, the visceral and proprioception, to the neuronal. The never-ending intertwinings and interactions between the fluctuating forces and intensities generated by these parallel psycho-physical systems are the material sources of Deleuzian 'sensation', and lie at the heart of my installations. This deeper mode of pre-reflective consciousness underpins observed or analysed sensation.

That one cannot consciously 'feel' the activity of these systems does not mean that these modes of consciousness do not have a role to play in our understanding of the more subtle aspects of our being. Rather the 'feelings' they initiate are 'closer to the inner core of life, and their target is more internal than external [with] profiles of the internal milieu and viscera playing a lead' (Damasio, 1994: 8) and are a necessary precursor to more conscious understandings experienced as sensation. Indeed, Damasio (1994) and Edelman (Edelman and Tononi, 2001) argue that the multiple dynamic modalities of the body (e.g. the processes of regulating life such as the nervous system, proprioception, temperature) induce the body states which underpin and permeate our reflective consciousness.

[C]ertain conditions of internal states engendered by ongoing physiological processes or by the organism's interaction with the environment, or both, cause responses which . . . allow us to have, amongst others, the background feelings of tension or relaxation, of fatigue or energy, of well-being or malaise, of anticipation or dread. (Damasio, 1994: 52)

Now, the interface between 'body' and technology in *Sensuous Geographies* (the video tracking system) had nothing to do with physiology, as it did not take directly into account the actual state of the participant's physiology that was elicited by the installation experience. Nevertheless, from an experiential perspective the physiological responses of each individual participant were an important factor in their behavioural response to the installation. The participants' verbal observations concerning their behaviour indicated that the autonomic, or at the very least semi-autonomic, physiological responses being elicited by the installation environment were guiding some of their motor responses. For example, when the underlying sonic context changed to a deeper sound, one participant noted that 'I just found myself going towards the floor, I just felt that I couldn't do anything else.' Others rose up on their toes, or their upper torso rose slightly and took on a more open quality when the sound environment became more 'open' and 'bright'. These kinds of responses were not responses to cognitive decisions to act in a particular manner, rather they emerged from the physiological systems which drive the 'undermind'.

Although, the interface between body and technology in *Sensuous Geographies* was not built into the computer system, as it is in biometrically driven installations such as *Whispers* (2002–2005),[17] it is ineliminably a central element on the ecosystem which underpins it, and a central part of the interactive dialogue between installation and player. Albeit non-wittingly, the physiological responses of many of the players in *Sensuous Geographies* to the sonic environment were as much a guide to behaviour as their conscious awareness of the effect their motion was having on the sounds. Thus the dynamic systems which underpin human activity need not be physically connected to an interactive installation system to become a part of that system, for they guide our behaviours without our knowing it.

Installations such as *Passing Phases*, *Time & Tide* and *Sensuous Geographies*, through the interface between the body and technology, are intended to draw attention to the more subtle aspects of our being in the world, and in doing so implicitly to sensitise participants to the 'hidden' sensibilities inherent in their responses to the environments in which they find themselves immersed, and to their dialogue with others.

Notes

1. Although in this chapter I refer to 'my' work and works 'I' made, all 'my' installations have been made in collaboration with other artists. In this chapter the personal pronoun incorporates the collaborative teams.
2. That said some installations, although primarily experiential, simultaneously offer players the opportunity to step outside the experience, and to view other players in the act of experiencing these liminal sensations. In many, even whilst viewing others, the player is immersed in the fluctuating audio-visual world, thus are implicated viewers.
3. An exception to this was Char Davies's work (www.immersence.com), which was beginning to introduce a degree of sensuousness into computer graphics, and was using physiological systems (notable respiration) as the interface between the body and technology.
4. The same set of images was available on each monitor, although as they were randomly selected the same progression of images did not occur on each monitor. See www.sensedigital.co.uk – *artworks* link.
5. See www.sensedigital.co.uk – *artworks* link.
6. A variety of video processing techniques were used to render the imagery abstract.
7. All quotations in this section are taken from the *Time & Tide* Visitor's book.
8. An interactive installation is more accurately described as a work-system than a work, as without a complex computer system the play of imagery cannot be brought to presence. Thus to some degree, it is the computer system that constitutes the 'being' of the work.
9. *Sensuous Geographies* is composed of 80 sound fragments and 77 independent abstracted images of the human body in motion. The sound fragments, and to a lesser extent the images, are open to modulation by the behaviour of up to four players (or four pairs of players). Five processing strategies (e.g. volume, pitch, reverberation) modulate the sound in accordance with the direction and velocity of players' trajectories as they move from one position to another. The composite behaviour of the individuals necessarily generates group behaviours.

10. In Boolean logic each programmed instruction can respond in one way *or* another ('if x then *either* y or z' *or* 'if x then y *and* z') depending on the conditions which obtain. Fuzzy logic implements variables on a continuous range of truth values, allowing intermediate values to be defined *between* the conventional binary (a 5'9' woman might be classified as both 'tall' and 'medium height').

11. An 'invisible player' sits at the computer and selects the sound samples and processing parameters, thus setting up the conditions for the musical structures of the sonic environment. He plays no further part in the modulation process.

12. Alternatively visitors can elect to stay outside of this inner space and observe the action as it takes place.

13. All comments are taken from participants who made entries in the *Sensuous Geographies* visitors book. www.sensuousgeographies.co.uk/

14. See www.sensuousgeographies.co.uk and www.sensedigital.co.uk – *artworks* link.

15. Andrew Benjamin argues that the work of art is '. . . not the object itself. . . but the continual questioning of the object. . . . the sustained presence of the work. . . ' (Benjamin, A. [1994]. *Object.Painting* London: Academy Editions, p. 17). Permeated by forces, even a painting or sculpture is, he argues, a becoming-object, an object of process.

16. It might also be seen as occupying the plane of what Damasio (2000) calls 'background emotion', that is the background feelings of tension, fatigue, well-being, anticipation which are frequently present without an identifiable reason.

17. Thecla Schiphorst, Susan Kozel *et al.* (2002–2005) http://whisper.surrey. sfu.ca.

References

Bergson, H. (1910 [trans.] first published 1899) *Time and Free Will: An Essay on the Immediate Data of Consciousness*. London: George Allen & Unwin.

Claxton, G. (1997) *Hare Brain, Tortoise Mind*. London: Fourth Estate.

Damasio, A. (1994) *Descartes' Error: Emotion, Reason, and the Human Brain*. New York: Avon Books.

Damasio, A. (2000) *The Feeling of What Happens*. London: Vintage.

Damasio, A. (2001) *Looking for Spinoza*. London: Vintage.

Deleuze, G. and F. Guattari. (1987) *A Thousand Plateaus: Capitalism and Schizophreni*. Minneapolis, London: University of Minnesota Press.

Deleuze, G. and F. Guattari. (1994) *What is Philosophy?* London: Verso.

Edelman, G.M. and G. Tononi. (2001) *Consciousness: When Matter Becomes Imagination*. London: Penguin.

Gallese, V., Fadiga, L., Fogassi, L. and G. Rizzolatti. (1996) 'Action recognition in the premotor cortex'. *Brain*, 119: 593–609.

Graham, M. (1991) *Blood Memory*. New York: Doubleday.

Kozel, S. and T. Schiphorst. (2002–2005) *Whispers*, Interactive Installation Project [Shown as *Whispers* Future Physical, Cambridge, UK 2003; *Between Bodies* OFFF Festival, Bilbao, Spain, 2004; *exhale: breath between bodies* Siggraph 2005, USA] http://whisper.surrey.sfu.ca/

Martin, J. (1939) *Introduction to the Dance*. New York: W.W. Norton & Co.

Massumi, B. (2002) *Parables for the Virtual: Movement, Affect and Sensation*. London, Durham: Duke University Press.

Morse, M. (1997) 'Video Installation Art: The Body, the Image and the Space- in between', in D.H. Hall and S. Fifer (eds) *Illuminating Video*. New York: Aperture/BAVC: 154–167.

Pelli, D.G. (1997) 'Two Stages of Perception'. Paper, *Theories of Vision* (Symposium) New York University. www.psych.nyu.edu/pelli/docs/pelli1997 symposium.pdf (accessed on November 2005).

Rodoway, P. (1994) *Sensuous Geographies: Body, Sense and Place*. London: Routledge.

Rubidge, S. and J. Rees. (2001) *Time & Tide*, Immersive Digital Installation [Presented in Chichester Fringe Festival, UK 2001] www.sensedigital.co.uk – 'artworks' link.

Rubidge, S. and A. MacDonald. (2003) *Sensuous Geographies*, Interactive Installation [Presented in New Territories Festival, Glasgow, UK 2003; Redcat Theatre, Los Angeles, USA 2004; The Showroom, University of Chichester, UK 2004; The Performance Gym, University of Winchester, UK 2005; Bowen West Theatre, Bedford, UK 2006] www.sensedigital.co.uk – 'artworks' link and www.sensuousgeographies.co.uk/

Rubidge, S., Hill, G., Diggins, T. and N. Parry. (1995–1999) *Passing Phases*, Interactive Digital Installation [Presented at Split Screen, University College Chichester, UK 1996; Place Theatre, London, UK 1997; IDAT, Phoenix Arixona, USA 1999] www.sensedigital.co.uk – 'artworks' link.

10
Body Waves Sound Waves: Optik Live Sound and Performance

Barry Edwards and Ben Jarlett

Director's perspective

My work with Optik falls into two distinct cycles of work. The first cycle developed over the period 1981–1986, and the second began in 1991.

Music and sound have always been key ingredients in Optik. From 1981–1986 the Optik sound was mainly sung ballads, and live acoustic solo instruments, the exotic collection of Optik's then resident musician Clive Bell – saz, accordion, shakuhachi, khene, and flute. These first-cycle performances were highly structured, colliding the unexpected with the familiar, the formal with the everyday, the spoken word with gesture. No part of the performance was ever improvised.

From 1992 to 2000 the company again used specific acoustic sources, mainly percussion (drums, cymbals, marimba and other instruments). This time, ambient or found sound also played an important role (footsteps, breathing, laughter and other sound), and more crucially these second-cycle performances were created making extensive use of improvisation.

Optik remained acoustic throughout these seven years (with one exception), until work took place with an electronic sound score for an Internet experiment in 2000. In the Autumn of that year the company gave a series of performances in Brazil (*In the Presence of People* Optik, 2000), and Optik's percussion player (Simon Edgoose) was persuaded to attempt an electronic experiment. For this project Simon Edgoose was in a London studio from midnight until early morning with an audience of 50 or so. The rest of Optik were performing live that evening in Sao Paulo. The sound was transmitted via an internet link from London to Sao Paulo and back again. There was also a video link but as this was coming in at around a frame every minute it bore little relation to what was happening in real time. The sound, however, came through with an array of digital delay, distortion and breaks of sequence. It worked. That experiment led to a whole new element in Optik performance: live electronic sound.

Two ways of making performance

The first cycle of Optik work did use electronically generated sound for one production (*Second Spectacle* Optik, 1982). I worked with the Electronic Music Studio of Keele University to engineer a dense rhythmic score, made of one long and one short sound repeated for the duration of the composition – about 12 minutes. Unmistakable Kraftwerk influence, in particular, their piece 'uno, duo, tres... quattro'. No laptops in those days, but a very large mainframe computer and lots of tape. This electronic segment was then slotted into the sound score as a whole, and accompanied the hanging of four black tail coats on clear fishing line, which then swung in the space, dancing somehow to the stylised and relentless beat. This use of layering and juxtaposition became a hallmark of my directing and creative style in this period. My technique and its performance results are described by Stella Hall in her review of Optik's third piece – *Short Sighted* (Optik, 1982):

There are images of extreme beauty and of sheer lunacy and sometimes they are the same. An eclectic assortment of instruments and electric singing combine to create a whole other dimension to the spectacle, and musician. The musician as performer in his/her own right means that the music doesn't simply become a theatrical backing track. The piece opens determinedly in the present. Heather Ackroyd in black plastic mac frenziedly attempts to break out of the mould, dancing a kind of T'ai Chi ballet as though her life depends on it. A boy in a red jersey joins her and is gone. Pan pipes play and a woman sings 'When lemons taste like honey dew, I'll stop loving you.' Before there's even time to register, time shifts and a becloaked figure with a lamp is serenading an unseen love. In the distance the woman sings: 'Don't play a sentimental melody / They leave me breathless / They leave me breathless.' The moments build layer on layer, each meeting oddly out of joint, as though two jigsaws had been shattered, then lovingly reassembled to form one. The count is advised by Figaro to disguise himself as a soldier in order to gain admittance to his loved one, he reappears as a soldier, but in modern dress. Rosina, as she flutters and flirts with the Barber, cannot see him with his sandwiches and thermos.

The action shifts, the elusive Rosina is found again on the other side of the world, a white-suited traveller finds her in some dusty shanty town where she dances a wild, shrieking display, a cigar clamped firmly between her teeth. A long slim flute is played, drums sound and, as she whirls, it seems the time is suspended. Optik excel at such moments, where the visual and the aural experiences meld and boundaries divide.[1]

Post the political events of 1989, I left what I felt was becoming a postmodern mainstream within performance to take a radically different direction towards creative process. I abandoned the collaging and precise structuring of Optik's first period to re-engage with the creative territory of an earlier project 'The Ritual Theatre' (1971–1974). Crucially, I began working again with spontaneity, with the unpredictable, in a word, with improvisation. There was the same desire to clash familiar with strange, and to collide the unexpected with the predictable, but in this approach the inner performance structure was of an entirely different nature. I worked with a series of actional sequences as before, but rather than pre-determine long passages of performance action I worked only on very small moment by moment actions, and relied on the decision-making processes of each performer in the work itself to create the final sequence(s).

The first approach could be described as working with blocks of performance material, the second cycle worked with wave-like transitions and emergent moments of event and action. The techniques I developed in this phase are described by curator–analyst Tracey Warr, written after she observed workshops and performances during the company's tour to Brazil (Optik, 2000):

Optik explore moving. And

> walking
> running
> colliding
> rocking
> falling
> rolling
> laying
> standing

sitting
seeing
looking
listening
feeling
focusing
waiting
deciding
being
stopping.

Walking. Taking a line for a walk. The three Optik performers walk and run in straight lines in dance studios in Sao Paulo and Campinhas. Their lines are moving sculpture in space. They make fleeting connections and collaborations. They fill a space, a void, gaps, with their moving. They fall into entrainment – walking or running together. They mirror each other. They lay down on the spot where someone else has just stood up. They walk to an internal rhythm – a body clock. They invade or do not invade invisible territories – body space, in your face.[2]

Optik's second cycle practice is based on the dynamics of the human body in space and time. It is inside this framework that the sound artist operates, whether acoustic or electronic. In other words the sound artist works as a performer along with the others, following the same creative decision-making process – not 'bolting on' pre-determined sound in a separate way.

We are all used to the notion of process in rehearsal and training. I was interested in how improvisation enables artists to take that process into the performance event itself. To work on how change occurs in performance, sudden change, unpredictable change, but ultimately any change at all. Performance is a complex process, and often performer techniques result in closing this down rather than exploring it. To work inside a complex process – essentially what ensemble creativity is all about – implies uncertainty and working with change that is outside of your control.

Performances are always a mix of what is put into a space, and what is already there. The size, shape and acoustics of a space, and of the performers and spectators in it, are all ingredients in the starting mix. Part of the pleasure of live performance lies in the way that the performers are drawing attention to themselves, in that space in that moment. Through this very human act the performer is both unique and connected to the watchers in some way. This is not something you can work out. As the phenomenologist would put it, it appears in its totality in one single act of intuition. This sense of connectedness or oneness can be explored in performance as a constant play/opposition with the sense of singular, uniqueness that each one of us is. As spectators and performers, we have an intuitive knowledge of this, of the essence of this connection.

We can experience sudden and total intuitive insights. Sometimes it feels as if the everyday crashes into the performance, full of unique particular instances. Some are audible: dogs, babies, drunks, coughs, police sirens, traffic, mobile phones, birds. Some are visible: individuals in the space, their belongings, bags, coats, the space itself, its doors, stairs, windows, lights. In these second-cycle Optik performances everything is made to seem ludicrously *in the way*, disconnected. And yet there is connection everywhere. The results of this intuitive understanding can surface as thoughts, feelings, perceptions or as action itself. What matters is this duality of particular and essence. The play between structure and chaos in performance demonstrates that theatre form and process are dynamic and can never be fixed. Theatre is a process of flux. I discovered a useful term to describe how this flux works in performer and spectator processes: 'eidetic intuition'. The word 'eidetic' comes from the Greek word *eidos* meaning form. It refers to a mental image that has unusual vividness and detail. Intuition is the immediate apprehension of something without reasoning. Eidetic intuition describes the immediate knowledge of the form of something, something that is vivid, and present. What theatre can do at its most fundamental human level is work with the material physical dynamics of the performer–space–spectator relationship to produce transformations in the human form itself.

Movement and stillness

There is a fundamental division between movement and stillness, silence and sound. It is the basic building block. To work with these basic blocks is to work on what happens *before* you make an action, what the French anthropologist Marcel Jousse termed *dynamogenesis*. This impulse into action, which results in all movement and sound, is an all or nothing thing. You can leave the build-up and go straight for the action, with the result that sudden, immediate, seemingly out of the blue change and transitions are possible.

This approach has its paradoxes and conundrums. To initiate a move I say to the performer 'do the action if / when you want to'. This is because you (the director) cannot give the instruction to do a particular action, since the instruction would become the cause, not the performer's own impulse, urge to act. All that matters is whether you act or not. There is no why. It is the action that is important and it is absolutely precise, nothing uncertain about it at all. You either act, or you remain still, silent. Two clear states, with no ambiguity:

> What Edwards achieves is the establishment of a complex and fragile 'enabling structure' to performance. 'Performance' emerges as an exploration. Edwards probes those aspects of human difference which become visible / audible when performers are asked and required constantly to make decisions about actional choices within a space and located before an audience (with all that this entails) . . . Edwards' practice is fluid, not fixed. His work opens itself up to the singular contributions of individual performers and is then able to explore this performer-singularity itself. He also makes available to each performer, in performance conditions, the space and the responsibility for split second decision making processes which bind each performer into the wider human and material structure of the performance event itself.[3]

As most performers know very well, the resolution of an action or sound sequence is usually the pay-off point, where the audience 'gets it'. But it is always more intriguing if the resolution of the impulse is left as potential for as long as possible. This is because without a rigid pre-determined resolution the movement can stand out as movement, the sound as sound, audible, visible, excessive, and the performers (the source of it all) with it. Causation and the need for it slips more into the background. Instead of asking 'why do I do this?' it is more productive creatively for the performer to explore the question 'shall I do this action now, or later?' No why (which is ultimately a matter of discourse). Instead there is only *when* (which is always precise, an event, not open to debate).

In both music and action, repeating is a key technical building block for the Optik process. Let us say you walk across the space. You reach the other side. You turn. You walk across the space. You turn. You walk across the space. Repeating in this way is a paradox (as chaos theory explains). Each move across the space feels like a repeat, but it is not identical in every respect to the move that preceded it. There are always minute differences. These minute differences can grow. In this way the significance of a moment, the emergence of an *event*, can arise not from the qualities of the movement itself, though certain qualities such as commitment and engagement are crucial, but from the conditions surrounding it. As these conditions are often outside the control of the performer, it follows that the performer cannot fully control what is significant and what seemingly is not. This is a strange but ultimately liberating understanding for a performer to grasp.

The separation of sound and vision into separate sensory processes for performance opens up many creative possibilities. As director I can ask performers to use the eyes to look at what they can see, but not to use the look to lead or direct the action. The emphasis shifts from the eyes and on to the whole sensory body. When a performer does this they work with what can be called corporeal perception. The focus of the eyes is then on whatever they can see in front of them, which will depend on the position of the body. Looking is tied into the body. But it is more than that. The shamanic song has the line 'my body is all eyes'. The performer becomes aware of the whole body as a sensitive system, based on the binary symmetry of the human body – forward-looking eyes, two legs, spinal column at the back, a front, and shoulders and ears that are to the right and the left side of the body. The performer uses this symmetry to know where they are in relation to everything that is outside of them. It gives them a sense of position. But again, it is more than just that. It develops a sense of *being there*, the human sense of proprioception. With this the performer can use the eyes not to look for something, but to look at what they can see. They can focus on a detail, or use the wider field of vision. We are able to do both. Movement is especially apparent at the edges of our field of vision. There is no need to turn the head as a result, which means that the performer can become aware of the precise line between seeing and not seeing. It is a very powerful line, and cuts across the line of forward movement.

Human hearing, however, works on a 360° basis. You can hear sound originating from a source that is out of sight. We are tuned to gauge distance from the acoustic dynamics of sound reaching us. Since we cannot turn our hearing off we hear everything. But not everything we hear registers with us in a conscious way (using the term loosely). Most sound in performance is used to support or reinforce the objective, whether it be the narrative or a key moment. I wanted the use of sound to go beyond the functional. The sound can seem to 'go with' a moment, or to 'go against' it. Actors can hear and respond actively to a sound, or work right through it, hearing but not responding (at least in any visible way).

Patterns of alignment in the space are formed from each performer's independent decisions about standing still, forward movement, rhythmic speed, turning right, turning left. The sound patterning and texture interacts with these spatial decisions, pushing, restraining, encouraging more speed, proposing a moment of stillness. The mathematician Poincaré said that there is no exact solution to all the possible movement relationships of three bodies in space.[4] Given this constantly changing set of possibilities there are endless different patterns, spatial and temporal, that emerge. But at any one time a performer can know what movement or sound pattern they are in, and the dynamics of its potential for change. In making these decisions each performer is constantly in the pull of the others. They have to be. If you are standing still and someone is running fast, just missing you, you will have the impulse to run with them, especially if the sound texture at that moment is with the runner and pushing them forwards. In these conditions we talk about resisting impulse or releasing impulse. You can do either. Its a kind of collaborate – compete situation. However, the essence of improvisation as a technique is that nothing stays the same for very long, change is constant, and wave-like.

Sudden realisation of the nature of a particular temporal or spatial relationship can often surprise performers with its intensity, leading to vocal and percussive sonic consequences. A performer, Hannah, was repeating a run across the space. The other performer, Simon, moved into her line and laid down on the floor. She continued to run, jumping over him when it was necessary. After repeating this action several times she started to laugh, and her laugh grew. Eventually she was running and jumping over Simon laughing and shrieking at the top of her voice. She had no warning about this sudden intensity that erupted in laughter. Or how long it might last. Physical sounds such as these coming from the actors' bodies, shrieks, laughter, breathing can be picked up by the live sound editing and woven into the sound pattern. This can produce strange fluctuations in the sense of time, of when something occurred (such as Hannah's final shriek). It is over in the live, but stays as a audible reality in the continuing sound-scape. Sound and action together work with these fluctuating emotional states. They can be described, to return to the philosophical, as a kind of phenomenological turbulence. As decisions are made live there is nothing that is predictable in their appearance or depth of feeling. In the language of new physics there are phenomena called *instantons* that are sudden and unpredictable eruptions of intense energy. They appear and disappear without warning. You could say that there are similar phenomena, similar

'instantons' of human experience, response and awareness in performance. Spectators are unavoidably (and necessarily) involved in these
fluctuations.

The arrival of digital sound

The sound in recent Optik work is always 'diagetic' (to use a cinematic
term). That is the source of the sound is always present, not imported.
This was why the imported pre-sampled voices, used in an earlier electronic experiment, did not work because the original sound source was
not present. It was for a tour of Poland in 1993 that Optik sampled a
soprano voice and wired different pitches of this voice, sung in long
notes, to a set of electronic drum pads. Each pad produced a particular
pitch and length of note. With six pads to play on, the drummer was
able to produce a stunning mix of harmonic and rhythmic sound, often
leading to textures of choral dimensions.

The system was used in the first performance on the Poland tour,
in the Rotunda Krakow, but was immediately abandoned because it
was too rigid and inflexible. The sound was great but unresponsive to
what was happening. The rest of the tour used only acoustic percussion
and sound. It was not until the interactive, responsive possibilities of
digital live composition arrived on the scene that I felt able to work with
electronic sound again.

Live sound and technology: Jarlett's system

I had been the sound engineer on Optik's earlier Brazil–London telepresence project, and soon after this Barry Edwards asked me to see if
I could find a way of generating an electronic sound score that did not
rely on using any ingredients produced in advance of the performance
itself. My solution to this problem was to use a granular synthesiser called
Granulab written by Rasmus Ekman.[5] In its initial state Granulab appears
simply to playback a sound file and loop it. I found that by using Granulab
in combination with a simple sound file recording and editing programme and an inexpensive boundary microphone placed in the space
he was able to create a sonic process that mirrored that of the performers.

Granular synthesis is a process first suggested by Iannis Xenakis (1971)
and Curtis Roads (1978) whereby sound is considered as a stream of
many small 'grains' of sound produced between several hundred and
several thousand times a second.[6] The idea originates from Dennis

Gabor's (1947) theory of sonic quanta, indivisible units of sound from a psychoacoustic point of view, which can be reversed without perceptual change in quality.[7] At its simplest this process separates the control of a sounds duration (time) from its pitch.

Looping

Once a recording is captured it is taken into Granulab and looped. The synthesiser's first controls define a loop start, length and rate. The start control defines where in the sound file to begin the loop, length defines how far into the remaining sound file to go before looping back to the start, and rate defines how quickly to move from the start to the loop point as a ratio. The rate control defines how fast the loop is played with respect to its original speed (forwards, backwards or held at a location in the loop). This control relates to that of the performer moving through space – a performer can move forwards, or stay still, change speed and direction. This looping control simply defines where in the sound file a grain when produced is to start from – in computational terms it could be described as a pointer.

Rhythm

The output of the grains of sound is independent of the loop described above. In granular synthesis, Barry Truax (1988) identifies that, for the ear to perceive, constant sound grains need to be less than 50 ms in duration.[8] These grains have an envelope shape (fade in, fade out), and frequency (how often they are produced – often measured as density). What I found to be interesting for the Optik project was to produce grains in a way that goes beyond this definition of granular synthesis – to produce grains with lengths that were perceivable (longer than 50 ms) and at low frequencies, producing rhythmic patterns from the recorded sound that evolve as they proceed. The sound produced happens regularly, giving a rhythmic pattern for the performer to react to, but its content can change with every repeat. Often in a recorded sound file used there are spaces in-between sounds, for example in a recording of footsteps a grain could coincide with the sound of a footstep on the recording, or in between them. This results in the rhythmic patterns having gaps, missed beats if you like, providing more interesting material for the performer to react to.

Texture

By altering the grain length, grain frequency and grain pitch (how fast a grain itself is played back as compared to the speed it was recorded) as well as other controls, I was able to create diverse timbres – from rumbling bass to high frequency distorted screeches. Moving from rhythmic patterns at various speeds to lush ambient drones became possible. These sounds, although abstract, fitted the performance because they were part of it, they came from it, and they influenced it.

Optik's first experiment in live digital sound took place in a studio of the Rambert Dance School in West London in 2001. I brought a simple mike, and two tiny speakers along with my computer. No one, including director Barry Edwards, had any idea what to expect. My first recordings picked up footsteps and some vocal sounds – laughter and breathing. What came out once I had worked on it astounded me, and everyone in that studio. While the performers were still moving, a sound score could be heard which had been derived minutes earlier from the actual sound of the performers themselves. Not only that, the performers could then respond to this sound, and I could respond back – nothing being fixed – meaning there was an improvisational dialogue between electronic sound and the actional sequences. Just what had been hoped for. After that things moved on fast to re-introduce acoustic music. I worked with voice, viola and wind instruments, for example. But the basic principle had been discovered, namely that you can have a responsive electronic sound score, composed moment by moment, and using only present sound sources (i.e. diagetic), which can work in live performance with actors. It was a real breakthrough.

I went on to practice and learn to play Granulab as an instrument during rehearsals – using a midi controller to control the software. I learned to improvise with it – finding its limits, familiarising myself with the control and process.

Granulab was first used in *takingbreath* in Belgrade (Optik, 2001). During this performance the sounds captured included not only those created by the performers and audience but also the sounds from outside the space added themselves into the mix. In one of the *takingbreath* performances, the squeaky vibrant sound, screeches and rumbles, of the trams outside Belgrade's vast SKC theatre hall were captured and worked into the sound score. Far from being a distraction, they were a welcome new source of material. By looking towards the windows, an actor could acknowledge the original source of the sound and, by carrying on without acknowledging, could be immersed in a strange vibrant soundscape that enveloped the action for no apparent reason.

What was to be the next breakthrough in sound had in fact presented itself by chance during London rehearsals for the Belgrade tour. One of the Optik performers, Simon Humm, ran into a telephone attached to the wall of the rehearsal space... the dialling tone was picked up... the rehearsal continued uninterrupted. Seconds later the dialling tone moved from a hum to a pulsating squeal demanding that the receiver be replaced. This was the first pitched noise to enter my system. Using Granulab T5 (an 8-channel version of Granulab) I was able to create layers of pitches – which he could then recycle through the system, from speakers to microphone through the space.

This new level to the sound score came up in a discussion Barry Edwards and I were having about pitch (over breakfast in Belgrade). The result was a decision to reintroduce acoustic musicianship into Optik performance. Initial experiments were done with voice as an input but this was superseded by the introduction of viola (Billy Currie) with the show *stream* performed at the ICA London (Optik, 2002).

Video performance was also introduced at this time with the addition of Howie Bailey who used VJ software (Resolume/Visual Jockey) mirroring my processes, using a camera and projector instead of a microphone and speakers.

I would work closely with Billy Currie over several months, improvising together, learning how to find combinations of feedback, harmony and silence. During this time I created new software systems using the graphical programming language MaxMSP[9] and granular synthesis externals written by Nathan Wolek.[10] My systems focused on reducing the time between when a sound was captured and when it could be used – until it was practically instant. An example of this was during Optik's performance *Space* (London 2004) when performer Alison Williams–Bailey used her voice to sing a sustained note. Using my new system I was able to put it out into the sound score before she had stopped. This would have been impossible to do using Granulab.

In the production *Xstasis* (Montreal 2003) I led, for the first time, an ensemble to mix the sound score. Four music students from Concordia University were taught to use granular synthesis software. Having a computer each and a selection of sound sources, they created individual granular textures on a computer which were fed into a computer, which I controlled, and out to four channels, creating a surround experience for the 18 performers and the audience present.

Artist–artist collaboration

Our collaboration as director and sound technologist has been based on our interaction as artists. Our aim has never been to create independently interactive software systems. In this respect our creative

Figure 14 Optik perform *Xstasis* (2003), Montreal, Canada (Photo: Alain Décarie)

Figure 15 Optik perform *Xstasis* (2003), Montreal, Canada (Photo: Alain Décarie)

practice is driven by the search for human content, human interest, the human story. In July 2004 a link was made with the first cycle of work when former Optik musician Clive Bell came to work with the current company. Clive Bell's speciality being wind instruments, in particular Eastern flutes, the sound score created was the result of the improvising musicianship on the Japanese shakuhachi and the processing musicianship on the laptop. Optik had brought together the most ancient of instruments and techniques with one of the most contemporary, exemplifying the artist-led nature of Optik's work.

Notes

1. Taken from Stella Hall. (1983). 'Optik Cockpit (Short Reviews)', *Performance Magazine – The Review of Live Art*, December/January, 20(21): 48–53.
2. See Warr, T. (2003). 'A Moving Meditation on a Dead Line', *Performance Research Journal*, December. London and New York: Routledge.
3. S. Melrose (1992). *Tank Research Paper.* Edwards and Melrose worked together on a week-long lab session with performers which resulted in the performance *Tank.*
4. Henri Poincaré (1854–1912), one of the first mathematicians to formulate the ideas of dynamical chaos.
5. Granulab can be found at http://hem.passagen.se/rasmuse/Granny.htm.
6. See I. Xenakis (1971). *Formalized Music: Thought and Mathematics in Composition.* Bloomington: Indiana University Press; and C. Roads (1978). 'Granular Synthesis of Sound', *Computer Music Journal* 2(2): 61–62.
7. See D. Gabor (1947). 'Acoustical Quanta and the Theory of Hearing', *Nature*, 159(4044): 591–594.
8. See B. Truax (1988). 'Real-Time Granular Synthesis with a Digital Signal Processor', *Computer Music Journal*, 12(2): 14–26.
9. Cycling 74's MaxMSP can be found at http://www.cycling74.com.
10. Nathan Wolek's Granular Toolkit can be found at http://www.nathanwolek.com/

References

Gabor, D. (1947). 'Acoustical Quanta and the Theory of Hearing', *Nature*, 159(4044): 591–594.
Roads, C. (1978). 'Granular Synthesis of Sound', *Computer Music Journal* 2(2): 61–62.
Truax, B. (1988). 'Real-Time Granular Synthesis with a Digital Signal Processor', *Computer Music Journal*, 12(2): 14–26.
Warr, T. (2003). 'A Moving Meditation on a Dead Line', *Performance Research Journal*, December. London and New York: Routledge.
Xenakis, I. (1971). *Formalized Music; Thought and Mathematics in Composition.* Bloomington: Indiana University Press.

11
Intelligence, Interaction, Reaction, and Performance

Susan Broadhurst

In this chapter, I will discuss my practice-based project, 'Intelligence, Interaction, Reaction and Performance', which consists of a series of performances that utilise new technologies. The first was *Blue Bloodshot Flowers* performed at Brunel University, West London, and the 291 Gallery, London, in 2001, and the second is *Dead East, Dead West*, which was performed at the Institute of Contemporary Arts, London, in August 2003. The performances which I directed consist of various physical/virtual interactions using a diverse range of technologies including motion capture,[1] artificial intelligence,[2] and/or 3D animation.

As a result of these technological advancements, I believe that new liminal spaces exist where there is a potential for a diverse creativity and experimentation.[3] These spaces are located on the 'threshold' of the physical and virtual, and as a result tensions exist. Since no body, not even a naked body, escapes representation altogether (Broadhurst, 1999: 103), the virtual body (as any other body) inscribes its presence and absence in the very act of its performance leaving gaps and spaces within its wake. I suggest it is within these tension-filled spaces that opportunities arise for new experimental forms and practices.

Blue Bloodshot Flowers was a collaboration with a computer scientist, Richard Bowden.[4] It is a movement-scripted piece. It was written by a colleague, Philip Stanier, and involves the remembrance of a love affair.[5] It is fairly ambiguous whether between an adult and child or between two adults, and even whether the characters are alive or dead. It is left to the audience to make their own meaning of who of what is what. The performance consisted of the real-time interaction between a human performer, Elodie Berland, and Jeremiah, an avatar (computer-generated image), and also between Jeremiah and the audience. Berland was French and we used a French voiceover as a memory device with

Figure 16   Elodie and Jeremiah from *Blue Bloodshot Flowers* (2001) (Image by Sally Trussler and Richard Bowden)

good effect. There was also some music used intermittently throughout, provided by David Bessell from the London College of Music.

Jeremiah is a head model based upon Geoface technology (DECface).[6] One of the most interesting aspects of this performance is how much the performer/spectator projects onto the avatar. This is not so surprising since a substantial area of the human brain is devoted to face recognition and the right non-dominant hemisphere of the brain takes a leading role in this (Zeman, 2002: 216). The ability of humans to recognise facial expressions is so sophisticated that even very slight differences are noticed and made meaningful. In this performance, it was remarkable just how much information could be gleaned by the spectators from Jeremiah's facial expressions on very little evidence, leading to a variety of emotions being projected by them onto the avatar.

Jeremiah consists of computerised artificial intelligence with the ability to track humans, objects, and other stimuli and to react to what's going on near him directly and in real time. However, interacting with Jeremiah is anything but objective. Most people, when they first see Jeremiah, find him 'spooky'. Then, after the initial contact leads to a degree of familiarity, people tend to treat him as they would a small child or a family pet. They usually try to make him smile and generally to please him. For instance, his face demonstrates sadness when he is left alone, so much so that many people find it difficult to walk away.

From a technological perspective, Jeremiah is based around two subsystems: a graphics system, which constitutes the head; and a vision system that allows him to see. The vision system surveys the scene via a wide-angle camera lens placed above the backdrop of the performance space and sends information to the head model which interacts with human objects or other stimuli. So Jeremiah is both the vision system and the head model. He also contains a simple emotion engine that allows him to respond to visual stimuli via expressions of emotions. The entire system is capable of running on a single PC but for speed of operation each subsystem ran on its own dedicated PC connected via a network crossover.

Jeremiah's head contains a simple Newtonian model of motion with random elements of movement, blinking, and ambient motion (Bowden *et al.*, 2002: 127). The Geoface-articulated bone model, DECface, provides a lifelike facial avatar that can be animated to produce various facial expressions. The software was custom written and produced by Bowden who 'prescribed' what Jeremiah's expressions would actually look like. Four basic pre-scripted expressions for key emotions are used within the system: happiness, sadness, anger, surprise (p. 125). Jeremiah's vision system is based around a Gaussian mixture model of colour distributions (statistical order of the colour of each pixel within an image) that uses expectation maximisation within the Grimson motion tracker framework.[7] This allows Jeremiah to probabilistically differentiate between the foreground and background pixels of a new image. The visual system additionally suppresses shadow and removes noise allowing static background scenes to be learned dynamically at the same time prioritising foreground objects (pp. 125–26). Jeremiah's attention is randomly distributed between these objects, weighted by their size and motion. Therefore, objects closer to Jeremiah appear larger and capture his attention more than objects further away, thus leading him to interact with the foreground objects in real time via expressions of emotions.

Although Jeremiah is programmed to react to certain stimuli with specific facial emotional expressions, he can also demonstrate random behaviour that can be disruptive during a performance. This unpredictability adds a further 'real life' dimension to working with a virtual being. This aspect of the performance questions orthodox notions of origin and identity since Jeremiah's identity is in no way fixed and his origins are not easy to specify beyond listing some technical specifications. As well as questioning conventions of authorship, ownership, and intertexuality, the digital technology that created Jeremiah subverts assumptions of reproduction and representation because in every performance Jeremiah is original, just as an improvising artist is original. Jeremiah is literally 'reproduced again' and not 'represented'.

Blue Bloodshot Flowers was divided into two sections. The first part consisted of a scripted movement-based interactive piece with Berland, while the second part involves spectators who were invited to interact directly with Jeremiah and to explore his supporting technology. Surprisingly enough, in the first part of the performance, although initial interest and curiosity were directed towards Jeremiah, the spectators' attention was mainly focused on Berland. However, the spectators' focus shifted to Jeremiah when he decided to display fairly inappropriate behaviour such as demonstrating happiness at an intense moment in the performance. We had no way of controlling his behaviour, which he learned as he went along. We could, of course, turn him off but we were very reluctant to do this. Jeremiah was the sole focus during the second part of the performance when he directly interacted with the spectators. Because I had decided not to restrict entrance to the 291 Gallery, audience members arrived right up until the very end of the scripted performance. I allowed unrestricted entrance for the very reason that Jeremiah would interact with any new arrivals he spotted. And of course he did, which amused everyone, except possibly the late arrivals.

Dead East, Dead West, the second performance in the series, was an experimental sound- and movement-based piece with some fragmented script, and was fused with 3D interactive technology. It was a collaboration between a choreographer from the Laban Centre, London, Jeffrey Longstaff; digital interactive artists from the University of the West of England, Martin Dupras, Jez Hattosh-Nemeth, and Paul Verity Smith; and an independent 3D filmmaker, Brian McClave. It consisted of motion tracking, interactive pads, and miniature cameras worn by the performers, Katsura Isobe and Tom Wilton, and also by the composer and percussionist, Dave Smith, and it was filmed and projected in real time in 3D. When seeing the performance with polarised glasses the captured images come straight out at the spectators and the performers appear to be moving through the images if viewed front on. Unfortunately this is lost in video clips of the performance and the image looks slightly distorted and unclear.[8] For the performance we used several cameras and at least eight projectors. The performers were captured and appeared in visual patterns on the rear screen in real time as well as being filmed and projected in 3D. The interactive pads, which were located around the space, produced various visual and aural effects. The images from the miniature cameras were projected on the side screen. The results of the miniature cameras which were worn by two of the performers on their heads are perhaps unique to the singular performer/audience perspective they provide the audience and each other with images of what they directly see, giving a new dimension to interactive performance. In addition, *Dead East, Dead West* was an intercultural and interracial performance that explored and exploded the margin between what is seen as dominant Western art practices and the 'exotic' performance of the 'other'.

Figure 17 Katsura Isobe, Dave Smith and Tom Wilton in *Dead East, Dead West* (2003a) at the ICA, London (Image by Terence Tiernan)

This performance was created as a collaborative exercise. Everyone involved exchanged ideas and experimented with modes of interaction in order to achieve a full and balanced integration of the generative and interactive content with the live performance. This was important since one key objective was to avoid the multimedia content appearing to have been added on as an afterthought. Moreover, the range of possibilities was known beforehand which made the dialogue and exchanges of ideas both challenging and exciting.

Most of the generative work was enabled with the open-source free software Pd (Pure Data) written by Miller Puckette.[9] Pd is a real-time software programming environment for the live performance of music and multimedia. It allows a programmer to create a series of patches for a specific purpose, rather than depending on an existing programme which carries large computing overloads; therefore, making the computer's response as fast as possible. Graphics Environment for Multimedia (GEM), a sub-programme, purpose written by Mark Danks that runs under Pd and provides real-time image manipulation, was also used. And Macromedia Director was used to develop the rest of the work. The software was chosen not only for its robustness but also for the ease of programming changes in real-time during performance (Dupras *et al.*, 2003: 2).

The 3D image processing was realised using two computers, one for sound input analysis and sending network messages over a local area network (LAN) to the second computer which was receiving live video input from a DV camera, and displaying the 3D visuals using GEM. Motion detection was used on this computer to deduce the amount of movement and direction of the performers. It was also used to control multiple 3D objects in a virtual 3D space onto which the input video was mapped as a live texture. Several aspects of the 3D objects were organised and controlled by the data received from the motion detection, such as size, rotation, colour, and spatial organisation. The 3D virtual 'set' was then displayed with two LCD projectors through polarising filters on a specially constructed silver screen, which also displayed stereoscopic video of the performance seen from varying perspectives

According to Dupras *et al.*:

In all our experiments we have been searching to create theatrical experiences which use the computer and computer generated imagery in a manner which goes beyond a simple 'stage-hand' effect. A more appropriate analogy would be that we are attempting to create instruments with which the musicians, dancers etc. can perform.

They are ideally however, more than simply a passive instrument in that they have an intelligence that will continue to act and react without cues from the performers or technicians. They are both another instrument and a performer in their own right. (2003: 2)

The live stereoscopic video processing and 3D image manipulation in *Dead East, Dead West* provided an ever shifting mise en scène through which the performance evolved. The imagery to a certain extent illustrated the emotions and the changing dynamics in the relationship between the two main physical performers. Finally, the entire performance including virtual environment and live performers was filmed and projected back onto the silver screen by McClave, again in 3D, which altogether created a richly textured, multi-layered, multi-media performance. The audience wore stereoscopic polarised glasses which allowed them to see the whole performance in 3D.

'Intelligence, Interaction, Reaction and Performance' is an ongoing project of what is hoped will be a variety of performances which combine the physical and virtual in performance. The rehearsal processes of both *Blue Bloodshot Flowers* and *Dead East, Dead West* proved extremely stimulating. Throughout, emphasis was placed more on the process of adaptation, how the performances developed and so on, rather than on the finished product. In this way, strategies are exposed and the apparent seamlessness of performance and technology is negated. Therefore, the objective is to destroy theatrical illusion, while at the same time resisting closure from within a place that is not completely aesthetic but is nevertheless performative.

The performances are hybridised and intertextual, and demonstrate such aesthetic features as heterogeneity, indeterminacy, reflexivity, fragmentation, a certain 'shift-shape style', and a repetitiveness which produces not sameness but difference. A central distinctive aesthetic trait is the utilisation of the latest digital technology. However, the digital as a discourse cannot convert phenomena directly but depends on a preceding production of meaning by the non-digital. Therefore, the avatar in *Blue Bloodshot Flowers* emulates the graphic design and animation of a recognisable representation, which is in this case a human head. The digital, like all formal systems, has no inherent semantics unless one is added. One must add meaning. Thus digitally processed contents require different than ordinary habits of reading – reading digital contents demands thinking in terms of 'indifferent differentiation' (Broadhurst, 1999: 177). A thinking that makes little distinction between the referent and meaning, or for that matter between 'reality' and representation.[10]

These performances can also be seen as critical deconstructive practices since 'metaphysical complicity' cannot be given up without also giving up the critique of the complicity that is being argued against (Derrida, 1978: 281). *Blue Bloodshot Flowers* and *Dead East, Dead West* are apparently complicit with dominant means of digital representation even as they try to destabilise those dominant structures. In other words, the performances address concerns regarding the commodification and consumerism of technology owned and provided by national and multinational corporations and used by the military.

The employment of wide, jarring metaphors is another central characteristic to these performances. The real-time interaction of the physical and virtual creates inclusive, jarring metaphors. This mixture produces an aesthetic effect caused by the interplay of various mental sense-impressions,[11] which 'unsettle the audience by frustrating their expectations of any simple interpretation and in so doing create a new kind of synaesthetic effect' (Broadhurst, 1999: 175). This is analogous to the experience caused by cross-wiring or cross-activation of discrete areas of the brain in certain perceptual disorders (Ramachandran and Hubbard, 2001: 9).[12]

Due to the hybridisation of the performances and the diversity of media employed, various intensities are at play. It is these imperceptible intensities that give rise to new modes of perception and consciousness. According to Deleuze and Guattari, 'experimentation has replaced all interpretation... No longer are there acts to explain, dreams or phantasies to interpret... instead there are colors and sound, becomings and intensities' (1999a: 162). Their view of art as 'sensation' – as a 'force' that ruptures everyday opinions and perceptions, 'to make perceptible the imperceptible forces' (1999b: 182) – provides a means of theorising the unpresentable or sublime of this kind of performance.

For Merleau-Ponty, instrumentation and technology is mutually implicated with the body in an epistemological sense. The body adapts and extends itself through external instruments. Giving the example of a blind man's stick, he claims that the stick becomes not merely a medium to locate the position of things but rather an extension of the man's own reach. Therefore, to have experience, to get used to an instrument, is to incorporate that instrument into the body (1962: 143). The experience of the corporeal schema is not fixed or delimited but extendable to the various tools and technologies that may be embodied. Our bodies are always open to and intertwined with the world. Therefore, technology implies a reconfiguration of our embodied experience. In these performances, the body is coupled with a variety of instrumental and technological devices that, instead of being separate from the body, become part of that body, at the same time altering and recreating its experience in the world. This intertwining of body, technology, and world is important since instead of abandoning or subjugating the physical body, instrumentation and technology extends it by altering and recreating its embodied experience. The body, in turn, creates new technologies and instrumentation to bring potential creativity and mediation into its corporeal world.

Moreover, the embodied self is central to both *Blue Bloodshot Flowers* and *Dead East, Dead West* where notions of the corporeal are prioritised. However, boundaries of this embodied self are not as fixed as we would like to believe. Since embodied emotional response is not only due to direct events that affect the self but can also be due to the stimulation of external objects that have been appropriated by the body and for that matter from the body (Ramachandran and Blakeslee, 1999: 61–62). Therefore, avatars such as those of Jeremiah, which are extensions and modifications of human beings, are likely to be emotionally appropriated by the spectator as another identity, becoming literally part of their own body.

In conclusion, although much interest is directed towards new technologies, it is my belief that technology's most important contribution to art is the enhancement and reconfiguration of an aesthetic creative potential which consists of interacting with and reacting to a physical body, not an abandonment of that body. For it is within these tension-filled liminal spaces of physical and virtual interface that opportunities arise for new experimental forms and practices.

Notes

1. 'Magnetic or optical motion capture has been used widely in performance and art practices for some time now. This involves the application of sensors or markers to the performer or artist's body. The movement of the body is captured and the resulting skeleton has animation applied to it. This data-projected image or avatar (Hindu: manifestation of a deity or spirit) then becomes some part of a performance or art practice' (Broadhurst, 2002: 159).
2. 'The consensus . . . is that AI is about the design of intelligent *agents*. An agent is an entity that can be understood as perceiving and acting on its environment. An agent is rational to the extent that it can be expected to achieve its goals, given the information available from its perceptual processes' (Jordan and Russell, 2001: lxxv).

3. 'Liminality', from *limen* (Latin: literally threshold) is a term most notably linked to Victor Turner who writes of a no-man's-land betwixt-and-between, a site of a 'fructile chaos . . . a storehouse of possibilities, not by any means a random assemblage but a striving after new forms' (Turner, 1990: 11–12). My own use of the term includes certain aesthetic features described by Turner, but emphasises the corporeal, technological, and chthonic (Greek: back to the earth) or primordial. Other quintessential features are heterogeneity, the experimental, and the marginalised. Therefore, *liminal* performance can be described as being located at the edge of what is possible (Broadhurst, 1999: 12).
4. Please see video clips and notes relating to the performance and the technology used (Bowden and Broadhurst, 2001).
5. See Philip Stanier (2001).
6. See Keith Waters (1987) and Waters *et al.* (1998), also Waters (1999–2004).
7. Jeremiah's vision depends on a background segmentation approach developed from an intelligent visual surveillance system based upon the work of Chris Stauffer and Eric Grimson. For more information, please see Richard Bowden *et al.* (2002: 25).
8. Please see selected video clips and notes relating to the performance and the technology used (Broadhurst, 2003b).
9. Puckette's new software system Pd provides the main features of Max/MSP and FST ('faster than sound') whilst addressing some of the shortcomings. According to Todd Winkler, 'by taking advantage of faster processing speeds, Pd is able to integrate audio synthesis and signal processing with video processing and 3D graphics in a single real-time software environment' (1999: 19).
10. For a more detailed discussion of the concepts of 'differentiation' and 'de differentiation', see Scott Lash (1990: 5–15).
11. 'Synaesthesia' is the subjective sensation of a sense other than the one being stimulated. For example, a sound may invoke sensations of colour.
12. This mingling of the senses presents in a variety of ways; for example, some individuals visualise see colours when they view numbers. Others see colours in response to a tone; for instance, a musical note may evoke a distinct colour. Various explanations have been given for this phenomena but recent evidence suggests that synaesthesia is a genuine perceptual phenomenon, not a high-level memory association (Ramachandran and Hubbard, 2001: 3).

References

Bowden, Richard, and Susan Broadhurst. (2001). *Interaction, Reaction and Performance*. Brunel University. http://www.brunel.ac.uk/jeremiah, accessed January 2006.
Bowden, Richard, Pakorn Kaewtrakulpong, and Martin Lewin. (2002). 'Jeremiah: The Face of Computer Vision'. *Smart Graphics*, 2nd International Symposium on Smart Graphics. Hawthorn, NY: ACM International Conference Proceedings Series: 124–8.

Broadhurst, Susan. (1999). *Liminal Acts: A Critical Overview of Contemporary Performance and Theory*. London: Cassell/New York: Continuum.

Broadhurst, Susan, dir. (2001). *Blue Bloodshot Flowers*. Performer: Elodie Berland. Music by David Bessell. Technology provided by Richard Bowden. Brunel University (June); The 291 Gallery, London (August).

Broadhurst, Susan. (2002). 'Blue Bloodshot Flowers: Interaction, Reaction and Performance'. *Digital Creativity* 13 (3): 157–63.

Broadhurst, Susan, dir. (2003a). *Dead East, Dead West*. Choreography: Jeffrey Longstaff. Performers: Katsura Isobe and Tom Wilton. Percussionist and composer: Dave Smith. Technology provided by Martin Dupras, Jez Hattosh-Nemeth, and Paul Verity Smith. 3D realization: Brian McClave (Filmmaker). London (August): Institute of Comtemporary Arts.

Broadhurst, Susan, dir. (2003b). Selected video clips and notes from *Dead East, Dead West. Body, Space, & Technology* 3, no. 2, Brunel University. http://www.brunel.ac.uk/bst, accessed January 2006.

Deleuze, Gilles and Felix Guattari. (1999a). *A Thousand Plateaus: Capitalism and Schizophrenia*. Trans. Brian Massumi. London: Athlone.

Deleuze, Gilles and Felix Guattari. (1999b). *What is Philosophy?* Trans. Graham Burchell and Hugh Tomlinson. London/New York: Verso.

Derrida, Jacques. (1978). 'Structure, Sign and Play in the Discourse of the Human Sciences'. In *Writing and Difference*. Trans. Alan Bass. Chicago: University of Chicago Press, 278–93.

Dupras, Martin, Jez Hattosh-Nemeth, and Paul Verity Smith. (2003). 'Beyond the Mechanical Stage-hand: Towards an Aesthetic of Real-Time Interaction between Musicians, Dancers and Performers and Generative Art in Live Performance'. In *Generative Arts: Proceedings of the 6th Conference*, ed. Celestino Soddu. DIAP Politecnico di Milano (10–13 December).

Jordan, Michael L. and Stuart Russell. (2001). 'Computational Intelligence'. In *The MIT Encyclopedia of the Cognitive Sciences*, ed. Robert Wilson and Frank Keil. Cambridge, MA: MIT Press, lxxiii–xc.

Lash, Scott. (1990). *Sociology of Postmodernism*. London: Routledge Press.

Merleau-Ponty, Maurice. (1962). *Phenomenology of Perception*. Trans. Colin Smith. London: Routledge.

Ramachandran, V.S. and Sandra Blakeslee. (1999). *Phantoms in the Brain*. New York: Quill.

Ramachandran, V.S. and Edward M. Hubbard. (2001). 'Synaesthesia: A Window into Perception, Thought and Language'. *Journal of Consciousness Studies* 8: 3–34.

Stanier, Philip. (2001). 'Blue Bloodshot Flowers: Text for Performance'. *Body, Space, & Technology* 1, no. 2. Brunel University http://www.brunel.ac.uk/bst, accessed January 2006.

Turner, Victor. (1990). 'Are There Universals of Performance in Myth, Ritual, and Drama?'. In *By Means of Performance*, ed. Richard Schechner and Willa Appel. Cambridge: Cambridge University Press, 1–18.

Waters, Keith. (1987). 'A Muscle Model for Animating Three-Dimensional Facial Expressions'. *Computer Graphics* 21, no. 4: 17–24.

Waters, Keith. (1999–2004). 'Decface'. Mediaport.net. http://www.mediaport.net/CP/CyberScience/BDD/fich_055.en.html, accessed Januuary 2006.

Waters, Keith, James Rehg, Maria Loughlin, Sing Bing Kang, and Demetri Terzopoulos. (1998). 'Visual Sensing of Humans for Active Public Interfaces'. In

Computer Vision for Human-Machine Interaction, ed. R. Cipolla and A. Pentland. Cambridge: Cambridge University Press, 83–96.

Winkler, Todd. (1999). *Composing Interactive Music: Techniques and Ideas Using Max*. Cambridge, MA: MIT Press.

Zeman, Adam. (2002). *Consciousness: A User's Guide*. London: Yale University Press.

12
The Tissue Culture and Art Project: The Semi-Living as Agents of Irony

Oron Catts and Ionat Zurr

This is an examination of the performative aspects of the Semi-Living (and objects of Partial Live) grown by the Tissue Culture & Art Project (TC&A). The TC&A, among a growing number of artists and collectives, is involved with the presentation of manipulated living systems in an artistic context. In contrast with art that deals with the representation of life through established artistic strategies, TC&A's type of engagement with living systems generate an experience which is closer to live/performance art. The phenomenological experience of the audience (as well as the artists) is of major importance for the TC&A. In much of TC&A's work the audience are 'forced' to actively participate in or be implicated with the alteration of the life cycle of problematised, technologically dependent fragments of life.

As part of the TC&A we look at Semi-Live Art (where humans and Semi-Living are unequally collaborating) as an attempt to challenge people's perceptions of life. Like performance or live art, TC&A is interested in the presentation of the subject rather than its representation. The audience is confronted by the existence of the Semi-Living through the different performative and aesthetic strategies of TC&A, but first and foremost by the fact that Semi-Living are sharing the time and space of the engaged audience. By presenting something that is 'sort of alive', that could only exist because of us and is dependent on us, TC&A lay bare the hypocrisies created to deal with the paradoxes in human relationships with other living beings.

What is TC&A? What are the Semi-Livings?

Since 1996, our artistic collective, The TC&A has questioned conventional notions of human relations with other living systems (whether human or non-human, living or partially living) and their anthropo-centric assumptions. This is done through the use of living tissues from complex organisms as a medium to create Semi-Living sculptures.[1] We are investigating our relationships with the different gradients of life through the construction/growth of a new class of object/being – that of the Semi-Living. These are parts of complex organisms which are sustained alive outside of the body and coerced to grow in predeter-mined shapes. These evocative objects are tangible examples that brings into question deep-rooted perceptions of life and identity, concept of self and the position of the human in regard to other living beings and the environment. We are interested in the new discourses, ethics, epistemologies and ontologies that surround issues of partial life and the contestable future scenarios they are offering us.

The Semi-Livings are constructed of living and non-living mater-ials; cells and/or tissues from, one or more, complex organisms grown over/into synthetic scaffolds and kept alive with an artificial support. The Semi-Livings are both similar and different from other human artefacts (homo sapiens' extended phenotype) such as constructed objects and selectively bred domestic plants and animals (both pets and husbandry). These entities consist of living biological systems that are artificially designed and need human and/or technological intervention in their construction, growth and maintenance.[2]

Semi-living and partial life can be seen as interchangeable terms. There are however some nuances; The Semi-Livings entities are usually shaped to forms that are not recognisable as being part of any Body in particular, partial life can be recognised as parts (i.e. an ear) of a whole of a living being. In the words of, one of France's leading philo-sophers and historians of science, Canguilhem: 'The introduction of cell theory in the biology first of plants (around 1825) and later of animals (around 1840) inevitably turned attention toward the problem of integrating elementary individualities and *partial life* forms into the totalizing individuality of an organism in its general life form.'[3] Symbol-ically, in the continuum of life, the Semi-Livings entities are nearer the non-living part of the scale, while objects of partial life are approaching the fully living.

Who is the performer? What is being performed?

In TC&A's work, we, the artists, position the Semi-Living entities on centre stage, while all the surroundings, including ourselves, become parts of the 'Techno-Scientific Body'. In the context of Semi-Living a Techno-Scientific Body is the artificial environment that sustain (and in some cases stimulate) the growth of living fragments of bodies. The Techno-Scientific Body includes the components such as a bioreactor, incubator, specialised nutrient solutions and other biological agents, as well as the human operators.

The TC&A draws on the Semi-Living as an attempt to subvert and develop a new discourse regarding the manipulation of living systems. However, this is difficult to transmit solely through live installations. TC&A, in most of our installations/performances, tries to shy away from portraying ourselves (including our own bodies) or the technology (which usually serves a utilitarian purpose rather than a fetishist one) as the focal point. Yet, in many cases the existence of the Semi-Living within the installations seems to be almost hidden by the bodies and technologies that already have a well-established contextual discourse. One can argue that the main reason for this is that the Semi-Living represent a condition/situation that lacks articulate cultural discourses and tools to respond to its existence, so most people will tend to ignore it, and focus on the familiar (in terms of both objects/subjects and discourse).

Furthermore, the Semi-Living purposely subvert binary positions such as human/animal, life/death, nature/culture as well as performer/performed. Another explanation might be that the Semi-Livings, though constantly changing, growing, mutating and dying, are doing so in a scale of time and space which is not easily detected by humans. All these dynamic processes of the life cycle are either too small or too slow to be absorbed instantly and can only be noticed over a period of time. We realised that even when we employ performative and interactive strategies, the audience who are faced with the Semi-Livings must exercise a leap of faith to believe that they are 'really' alive and not a hoax. Here the issues of performative space, time and scale are being stretched and played with: a durational performance of a Semi-Living can take months in a space that is smaller than a matchbox.

The Semi-Livings are dependent on artificial support for their survival. They are living fragments which were stripped of their 'host' body (whether living or recently deceased) and its immune system. As a result they have no way to resist infection when exposed to the external environment; they must be contained in sealed and sterilised vessels in order to survive. It means that every physical interaction with the Semi-Living is mediated through technology; in the form of a bioreactor, a pipette and a sterile hood. Furthermore, in order to maintain the life of the Semi-Living, we have to build a fully functioning tissue culture laboratory that provides the appropriate conditions and enables the procedures involved in caring for the Semi-Living. The presence of a laboratory and other technological apparatuses, we realised, made the audience so overwhelmed that it overshadowed the actual players in the show – the Semi-Living. In this context the Mise-en-scene, which is both functional and theatrical, takes precedent over the Semi-Living performer.

In our post-capitalist, techno-fetishist and anthropocentric society the performer is often viewed as either the technology or ourselves (who maintain the routine work in the laboratory), while the Semi-Living which are both process and outcome are seen as only a by-product. Needless to say, this phenomenon defies the whole notion of the TC&A project. Therefore we had to devise some ways of 'helping' the audience to 'escape' from their comfortable position of dealing with the work down the path of least resistance (resorting to known discourses). To move the audience into a realm where the subject/object on stage, although an assault to many of our pre-conceptions about life, death, identity and our environment (both natural and cultural) cannot be ignored, we developed the phenomenological rituals of Feeding and Killing.

TC&A's rituals – Performative and interactive strategies to deal with the Semi-Living

One way the viewer/participant can observe and appreciate the aliveness of the Semi-Living is by revisiting them over an extended period of time in order to see, with human eyes, the phenotypic changes. For those who cannot do so we devised our rituals.

The rituals are performed for practical reasons – maintaining the life and growth of the Semi-Living sculptures – as well as for conceptual reasons; by celebrating and terminating Semi-Living art forms, we trouble the conventional art viewer's autonomous reflective space (as does all performative art). Our installations involve performative elements that emphasise the responsibilities, as well as the intellectual and emotional impact, which results from manipulating and creating living systems as part of an artistic process.

The Feeding Ritual is performed routinely. Here we raise questions about the caring, tending and nurturing needed for all engineered and manipulated life forms, including Semi-Living sculptures/entities. We invite the audience to view the process of feeding which is done in a laboratory situated within the gallery as an integral part of the artistic experience. Initially we emulated the zoo-inspired ritual of the set feeding times. The audience have been informed about a set time every day in which the feeding will occur. That presented some problems for us; it seems that by linking (in one's mind) this activity to the familiar zoo ritual, people were expecting to see some tricks performed by the entity to be fed. When that did not happen, we, the feeders, become the spectacle. We decided to change it so that the feeding happened when it was needed rather than at an arbitrary set time. We found this strategy to be more effective as it represented more of a transgressive intervention into the way art is being seen in the gallery context.

As most of our exhibitions takes place in a visual arts context – our coming unannounced to tend to the Semi-Living made us more (in the eyes of the audience) of an unscheduled maintenance crew looking after something which is 'kind of alive'.[4] In this case the audience who happens to be present in the space at the feeding time is confronted with the 'liveness' of the Semi-Living – much less of a spectacle than in the case of the set feeding times. In many cases the dialogue which is formed between the performers/maintenance crew was also less formal and therefore more revealing than the more constructed routine of the feeding times.

We also shied away from the pristine white laboratory coat and its associated connotations and designed our own. It is a hybrid of a laboratory, chef and a mechanic coat and is grey in colour,[5] as our practice can be associated with different professions and locations such as the scientific lab, the domestic kitchen or the backyard garage.

In her thorough survey of the use of animals in art and entertainment, Thornton has divided the use of life in popular culture and as it is represented in art in this way: Living systems in Zoos and Menageries are represented as objects, in circuses and animal acts as performers, in sacrifice, factory farms and fighting as victims and in cultured pearls, honey bees, free range farms and so on as co-creators.[6] It is difficult to position the Semi-Living entities within these constructions; to a certain extent they are co-creators who we collaborate with. We have limited control over the eventual shape and fate of these entities. However, it is not an equal collaboration and we, the human artists, have much more power. They are also partly objects; can we say they perform or are they victims of circumstances which are forced upon them? Basically they are a collection of cells, outside of the context of their host bodies that people tend to anthropomorphise.

At the end of every installation, or in a situation when we cannot stay for the duration of the show, we are faced with the ultimate challenge of an artist – we have to literally kill our creations. The works have to be killed for practical and conceptual reasons. For this we devised the Killing Ritual. The killing is done by taking the Semi-Living sculptures out of their containment and letting the audience touch (and be touched by) the sculptures. The fungi and bacteria which exist in the air and on our hands are much more potent than the cells. As a result the cells get contaminated and die (some instantly and some over time).

The Killing Ritual enhances the idea of the temporality of life and living art, and our responsibility as manipulators of these new forms of life. The Killing Ritual can be seen as 'violent and pitiless act' of transforming the Semi-Living back to a 'sticky mess of lifeless bits of meat' (we as a society prefer to examine collection of cells disassociated from their organism, i.e. a steak in a butcher shop) or as an essential display of compassion; euthanasia of a living being that cannot care for itself and has no one to care for it.[7] We also make a point of inviting the people who invited us (curators, gallery directors etc.) to participate in the killing, as they also are responsible for the well being of the Semi-Living sculptures presented in their show. On more than one occasion people from the audience have approached us after the ritual and told us that only by killing our sculptures did they realise they were alive.

The *Pig Wings* Installation, Adelaide Biennale for Australian Art 2002

One interesting experience happened in the Art Gallery of South Australia as part of the Adelaide Biennale for Australian Art 2002, where we presented our *Pig Wings* Installation. The gallery management and staff knowing that we were presenting living tissues and constructing a functional wet laboratory in the gallery were worried and wary. The ironic title of our work – *Pig Wings* – was not very helpful. As a result of pressure on the gallery from the curators of the show, as well as our credentials as artists working in scientific institutions and constructing and operating laboratories in artistic and public places, we were able to build a small (claustrophobic) laboratory. The laboratory contained a sterile laminar flow cabinet, a small fridge that hosted the cells nutrient media and an incubator for the microgravity bioreactor that hosted the 'humble' Semi-Living *Pig Wings* constructs. These are three sets of wings made out of pigs mesenchymal cells (bone marrow stem cells) grown over/into biodegradable/bioabsorbable polymers (PGA, P4HB). The wings measure $4\,cm \times 2\,cm \times 0.5\,cm$ each and they were never intended to be implanted onto pigs.

As agreed with the gallery, we would come to perform the Feeding Ritual every day at a set time for 10 days. By the end of this period the *Pig Wings* would be killed and presented as dead relics for the rest of the duration of the show. Being a conservative art gallery, which contains only 'dead art', the staff were concerned about the fact that the work is alive, that it changes over time, and that we will be coming in every day to move it around (from the bioreactor to the sterile hood and back again). One can understand the anxiety they must felt having such a piece in their gallery. The added layer of a heavily scientific and technological overlay that was 'imposed' on the installation created a somewhat Frankensteinian fear that something will go 'horribly wrong'.

The gallery did not employ invigilators for the show but rather private security guards who were very nervous about our installation. Furthermore, they were not sure if looking after our laboratory was covered under their contract with the Gallery. After a private meeting with the relevant personnel we assured the staff that they were not allowed into the laboratory and should not perform any task associated with our installation including turning on/off computers and lights. Throughout the time that the *Pig Wings* were alive, these reluctant staff witnessed every Feeding Ritual and observed first hand the growth of the Semi-Livings. They were there when we talked about our fears that the *Pig Wings* will be contaminated, and realised the level of investment in time and emotion that goes into keeping them alive. When it was time to kill the wings, a couple of the security guards, who initially were very apprehensive, approached us and asked us to train them to look after the *Pig Wings* as they had grown attached to them and 'did not want them to die'. Unfortunately it was not possible to do so (mainly due to health and safety regulations) and the *Pig Wings* were killed.

Peter Sellers, the Artistic Director of the Adelaide Festival, was invited to participate in the Killing Ritual. Visibly emotional, wiping a tear from his eye he told us how each time he has to turn off the Bill Viola installation he feels how the artwork 'dies' (until the next day when the video projection is turned on again) but he never thought that he would literally kill an artwork just by touching it.

Extra Ear – ¹/₄ Scale, National Gallery of Victoria, Melbourne 2003

We had a very different though related experience with the partial life *Extra Ear – ¹/₄ Scale* (in collaboration with Stelarc). This was the first time we created an object of partial life that resembled a human organ – a $^1/_4$-scale-sized ear.

Extra Ear – $^1/_4$ Scale is about two collaborative concerns. The project presents a recognisable human part. It is being presented as partial life and brings into question notions of the wholeness of the body. It also confronts broader cultural perceptions of 'life' given our increasing ability to manipulate living systems. The TC&A deal with the ethical and perceptual issues stemming from the realisation that living tissue can be sustained, grown, and is able to function outside of the body.

We are interested in the ear as a stand-alone signifier of an independently existing part of the body, and less interested in the eventual attachment of the ear to the body. Even so it seems that this piece has managed to evoke reactions that none of our other works did. The religious view of the human body made in the image of God motivated some of these extreme reactions. While the State Gallery of South Australia (where we presented the *Pig Wings* Project) raised difficulties in regard to presenting living art, The National Gallery of Victoria (where we presented the ear in September 2003) was even stricter in its requirements. From a performative perspective there were three very significant issues that were sticking points in our attempt to realise the project in the context of this very traditional art venue. The first was their refusal to allow us to use human tissue for this installation. The second was their somewhat strange request for us to declare that the work does not raise ethical issues, and the third was their refusal to allow us to perform the feeding and killing rituals during opening hours.

According to the curators of the NGV, about two weeks before the show was about to open they realised that the gallery had no policy in regard to presenting living tissues. The Gallery Director instructed the curator to seek clarification in regard to the project including a statement from us that the work does not raise ethical issues in general and in the biomedical community in particular. We could not reassure the gallery that this is the case as we see the primary aim of our work to act as a tangible example of issues that need further ethical scrutiny, and to critically engage with the biomedical industry. This was stated as our aim when we applied for the human research ethics clearance from the University of Western Australia (UWA). Disregarding the fact that this installation received ethical, safety and health clearances from UWA, the NGV decided to cancel the installation, only to later 'compromise' and allow it to go ahead on the condition that we did not use human tissue. The compromise of using non-human animal cells, while keeping the proposition of the piece, enhanced the non-anthropocentric stand of the Semi-Livings and partial lives.

Our attempt to deal with the human form, in the context of a performative partial human life, received an interesting twist in our dealing with the art establishment. Much of the attention we received was a strong reaction against the disfigurement of the body with both the suggestion of implanting an ear onto Stelarc's body and the distinctively recognisable human body part (an ear). This seemed to trouble the NGV, as on a number of occasions they cited a previous controversial piece exhibited in 1997 in the gallery, 'Piss Christ' by Andrew Serrano. Much concern was about the possibility of the same religious group that vandalised Piss Christ and threatened staff attacking the *Extra Ear – ¹/₄ Scale*.[8] However, no such threat eventuated and it was only the NGV that made the connection between the *Extra Ear – ¹/₄ Scale* piece and the blasphemy by correlating perceived modification to the human form with disfiguring the image of God.

In addition, as mentioned above, the audiences were not allowed to see us tending to the ear. The notion of care and life was reduced to a little ear-shaped object floating in a reddish liquid in something that looks like a modified microwave oven. Even so, we received a couple of concerned phone calls from the gallery to inform us that the ear looks somewhat different from when they had seen it last and wondered if something had gone wrong. 'Such is life' we replied.

The victimless series – bringing the victim closer

Usually people who oppose our project find it difficult to articulate the source for their disapproval and react more from a knee-jerk impulse. We believe this is a result of the TC&A forcing people to reassess their perceptions of life by presenting life at its visceral and somewhat abject form as manifested by the Semi-Living.

All animal life is involved with exploitation of other lives in different degrees and for that humans have devised different mechanisms to be able to justify this exploitation ethically and morally.[9] In the Victimless series installations/performances we are dealing with these issues by destabilising the human power structure governing relationship between different living systems.

First, we considered the possibility of eating victimless meat by growing Semi-Living steaks from a biopsy taken from an animal while keeping the animal alive and healthy.

This artwork deals with one of the most common zones of interaction between humans and other living systems, and probes the apparent uneasiness people feel when someone 'messes' with their food. It also deals with one of the most hypocritical zones in the anthropocentric tyranny of humans on this planet. The project offers a form of symbolic 'victimless' meat consumption. As the cells from the biopsy proliferate, the 'steak' *in vitro* continues to grow and expand, while the source, the animal from which the cells were taken, is healing. Potentially this work presents a future in which the killing and suffering of animals destined for food consumption will be reduced. However, by making our food Semi-Living – we risk making the Semi-Living a new class for exploitation. In addition, the nutrients in which the steak is bathed contain animal-derived products. The distance from the victim sometimes makes us forget that almost any form of diet involves victims – no matter how processed, engineered or organic the food is.[10]

Our own research into this project began as part of our residency at the Tissue Engineering & Organ Fabrication Laboratory at Harvard Medical School in 2000. The first steak we grew was made out of pre-natal sheep

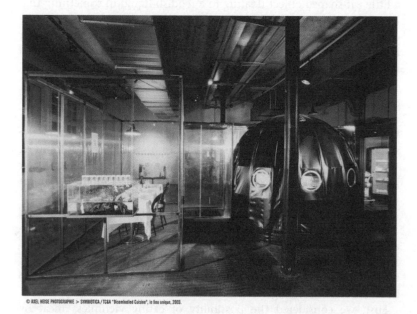

© AXEL HEISE PHOTOGRAPHIE > SYMBIOTICA / TC&A "Disembodied Cuisine", in lieu unique, 2003.

Figure 18   Disembodied Cuisine Installation, from L'Art Biotech Exhibition, France 2003, The Tissue Culture & Art Project 2003 (Photography: Axel Heise)

cells (skeletal muscle). We used cells harvested as part of research into tissue engineering techniques *in utero*. The steak was grown from an animal that was not yet born.

We finally were able to present and perform this project in 2003 as part of L'Art Biotech Exhibition in France. We titled the installation *Disembodied Cuisine*, playing on the notion of different cultural perceptions of what is edible and what is foul. We grew Semi-Living frog steaks, with the intention of raising questions about the French resentment towards engineered food and the objection of other cultures to the consumption of frogs. To our delight, we found a source of cells that did not require inflicting an injury to an animal. We ended up using an immortalised cell line (cell lines are either modified or cancerous cells that have the ability to grow and divide indefinitely and can be seen in the context of our work as a renewable resource). The cell line we used was developed using non-mutagenic techniques and according to the advice we received was considered safe for consumption. These cells were developed at a Japanese laboratory in the late 1980s from the skeletal muscle cells of a tadpole of an aquatic toad, *Xenopus laevis*. We grew these cells over a biopolymer scaffold for potential food consumption. Throughout the 3 months of the exhibition we tended to our steaks, feeding them every two days and protecting them from possible harm (like bacterial infection). On the last day of the exhibition we had the ultimate Nouvelle Cuisine style dinner to which members of the public and invited guests joined us to eat the two coin-sized steaks. Some of the eight diners found the steaks too hard to swallow.

Although the 'dinner' functioned as a Killing Ritual of the Semi-Living, it was also a symbol for the extension of life (as the animal from which the cells were taken continued to live). In Huxley's words: 'Not a necropolis, but histopolis... not a cemetery, but a place of eternal growth'.[11]

We collected the bits that were spat out for our follow up piece – *The Remains of Disembodied Cuisine*. We presented these spat out bits, on a dining table along side a three screen video installation that documented the project from the lab and farm through the exhibition to the final dinner.

In a later project in this series – *Victimless Leather* – we grew a miniature stitch-less jacket out of immortalised cell lines which formed a living layer of tissue supported by a biodegradable polymer matrix. The *Victimless Leather* project is concerned with growing living tissue into leather-like material. This artistically grown garment confronts people with the moral implications of wearing parts of dead animals for protective and aesthetic reasons and further confronts notions of relationships

Figure 19 Victimless Leather – *A Prototype of Stitch-less Jacket grown in a Techno-scientific 'Body'*, The Tissue Culture & Art Project 2004. Medium: Biodegradable polymer connective and bone cells (Dimension of original: variable)

with living systems manipulated or otherwise. An actualised possibility of wearing 'leather' without killing an animal is offered as a starting point for cultural discussion. Saying that however, the production of the 'leather' was not totally victimless – we still used animal-derived ingredients in the nutrients we provided the tissues with. Hence our reference to the 'victimless' is an ironic one, and is seen by us as a critique of the type of technological mediated promises of 'utopia'. This piece also presents an ambiguous and somewhat ironic critique of the technological price our society will need to pay for achieving 'a victimless utopia' as the stitchless jacket that was grown as part of this project could only survive within a Techno-Scientific Body – a bioreactor.

As part of the 'victimless series' two new projects are in the pipeline. These projects are about to take the victimless utopia to somewhat extreme levels of absurdity. *The DIY De-victimizer Kit Mark One (DVK m1)* is set up to allay some of the guilt people feel when they consume parts of dead animals (either for food, aesthetics or any other purpose) or cause an accidental death to a living being (by a car, a lawnmower or any other piece of technology). The kit can maintain and in some cases even proliferate and extend the life parts of the deceased bodies, at least until the guilt (stemmed from the accidental killing) recedes. The *DIY DVK* utilises off the shelf items to construct a basic tissue culture facility, some specialised nutrients are needed – some of which contain animal-derived material – but it is so far removed from the end user that for most people remorsefulness is not usually an issue.

Relifing Roadkill will make use of the *DIY DVK* for a performative installation in which we will experiment with bringing back to life (literally) parts of road kills. By 're-life-ing' the victims of our technological-fetishistic society, the TC&A will further explore our relations with other living systems, whether living or Semi-Living, whether human or not. What is then the status of life generated by something that we mostly take for dead? What is the point in relifing roadkill? This piece will let us take a somewhat ironic view on the interplay between the made and the biological, the constructed and the visceral, the simulation and the 'real' by extending parts of life of which our technological-fetishistic society is responsible for their premature death. *Relifing Roadkill* will explore the technological mediated victimless utopia through a phenomenological experiment with the effects of our technology – in this case our cars – over biological systems – road kills. We will attempt to partly reverse the 'destructive' effects of human technology by re-life-ing its parts of its victims. In this participatory project the audience will take an active role in the experiment by collecting the fresh corpses, assisting us in caring for the fragments of life and making different ethical decisions in regard to their eventual fate.

Conclusion

Artists can play a role in exploring these issues and *Spawn* 'philosophy in action'. There is a growing discrepancy between our cultural perceptions of life and what we know about life scientifically and what we can do with life technologically. Our work deals with the tension between caring for living systems on the one hand and instrumentalising life on the other. We believe that live art is best situated to confront such a paradox in ways that constructively raise philosophical and epistemological issues.

Henry Harris writes about the biological phenomenon of cell fusion which is described as 'violating the most fundamental myths of the last century':

> Any cell – man, animal, fish, fowl, or insect – given the chance and under the right conditions, brought into contact with any other cell, however foreign, will fuse with it. Cytoplasm will flow easily from one to the other, the nuclei will combine, and it will become, for a time anyway, a single cell with two complete, alien genomes, ready to dance, ready to multiply. It is a Chimera, a Griffon, a Sphinx, a Ganesha, a Peruvian God, a Ch'i-lin, an omen of good fortune, a wish for the world.[12]

Harris interpretation of this phenomenon may seem fairly positivist and somewhat romantic, however, it might be seen as a prelude to TC&A notions of the Extended Body: The living fragment becomes part of a different order that includes all living tissues regardless of their current site and they are cared for by the Techno-Scientific Body.

In many ways, the performative aspects of the TC&A artists (as opposed to the living performance of the Semi-Living entity) are rather the operation of the Techno-Scientific Body. Ironically, we have created the Semi-Livings to 'make' us perform, in front of the audience, our care for them. The excessive act of engineering a 'new sort of life' made us perform a ritual which emphasises our subordination to its needs/demands. These Semi-Livings have made us part of their techno scientific body – as much as they are the creations/extensions of our bodies, we are theirs.

There is an inherent paradox to the Semi-Living performers: Seen as yet another apparatus of society increasing instrumentalisation of life they are also ironical agents that reveal to us, humans, our (sometimes reluctant) submissiveness to the actuality of being an unseparated part of the greater life continuum.

Notes

1. The way we obtain our raw materials is important for us to note; we are scavenging leftovers from scientific research and/or food production.
2. For more, see O. Catts. and I. Zurr, 'Growing Semi-Living Sculptures', *Leonardo* 35:4. Massachusetts: MIT Publishing, 2002: 365–370.
3. G. Canguilhem, in Delaporte, Francois (ed.) *A Vital Rationalist: Selected Writings from George Canguilhem*. New York: Zone Books, 1994: 84–85.
4. It happened that security was called upon us, suspecting us as intruders.
5. The grey colour is referencing the laboratory where the first successful tissue culture experiments were performed in 1910. This lab hosted the 'Experimental Surgery' Group, led by Dr Alexis Carrel of The Rockefeller Institute for Medical Research. In a 1954 book titled *The Cultivation of Animal and Plant Cells*, P.R. White refers to Carrel's lab and the results of his practice:

   > The grey walls, black gowns, masks and hoods; the shining twisted glass and pulsating coloured fluids; the gleaming stainless steel, hidden steam jets, enclosed microscopes and huge witches' cauldrons of the 'great' laboratories of 'tissue culture' have led far too many persons to consider cell culture too abstruse, recondite and sacrosanct a field to be invaded by mere hoi polloi!

   Dr Carrel is a controversial figure in the history of Bio-Medical research. He was the first to develop tissue culture techniques. Yet he is considered an eccentric mystic and fascist (or at least a Vichy – collaborating eugenicist. For more, see TC&A paper in the *LÁrt Biotech* Exhibition Catalogue, le lieu unique France ISBN 2–914381–52-2. 2003: 20–32.
6. K.D. Thornton, 'The Aesthetics of Cruelty vs. the Aesthetics of Empathy', in *Aesthetic of Care*, ed. Oron Catts. SymbioticA Press, 2002: 15–21.
7. In *Art and Fear* Virilio use the terms 'pitiful' and 'pitiless' to describe living/presentative art. He associates 'pitiful' with compassionate and symbolic representational art and 'pitiless' with merciless, disfiguring presentative contemporary arts. For more, see P. Virilio, *Art and Fear*. Translated by Julie Rose. London & New York: Continuum, 2003.
8. For more, see http://www.artslaw.com.au/Publications/Articles/97Blasphemy.asp.
9. Beside the anomaly of the Symbiotic flatworms that as described by L. Margulis and D. Sagan 'are such good providers that the worms have atrophied mouths; the close-mouthed green worms "sunbathe" rather than seek food, and the endosymbiotic algae even recycle the worm's uric acid

waste into food' from *What is Life*. Berkeley & Los Angeles: University of California Press, 2000: 120.

10. As any farmer know any type of commercial and semi-commercial crop growing involves the killing or disturbing of animal life; form mice to worms and caterpillars as well as the plant life itself which has much more agency and order than the Semi-Living.

11. J. Huxley, 'The Tissue Culture King', in *Great Science Fiction by Scientists*, ed. Groff Conklin. New York: Collier Books, 1962: 158–159.

12. H. Harris, 'Roots: Cell fusion', *BioEssays* 2: 4. Wiley Periodicals, Inc. A Wiley Company, 1985: 176–179.

References

Catts, O. and I. Zurr. 'Growing Semi-Living Sculptures', *Leonardo* 35:4. Massachusetts: MIT Publishing, 2002: 365–370.

Huxley, J. 'The Tissue Culture King' in *Great Science Fiction by Scientists*, ed. Groff Conklin. New York: Collier Books, 1962: 158–159.

Virilio, P. *Art and Fear*. Translated by Julie Rose. London and New York: Continuum, 2003.

White, P.R. *The Cultivation of Animal and Plant Cells*. NewYork: Ronald Press, 1954.

13
Addenda, Phenomenology, Embodiment: Cyborgs and Disability Performance

Petra Kuppers

Disabled cyborgs

The cyborg is a person with addenda – something strange, foreign, other is added to the basic ingredients which denote 'human'. Cyborgs abound in popular culture, myth and legends, and in contemporary culture, the cyborg has a greater presence than ever. Many of the cyborgs of contemporary cinema, digital storytelling and visual culture are disabled people, enhanced and supplemented by technology. Some cyborg discourses show a fascination with the sensual and kinaesthetic experience of 'being-added-to', being different. In particular, some of these discourses which find expression in popular visual material focus on lived experience and the corporealisation of the cyborg – on the performance aspects of visual media. I want to argue that the sensationalised 'addenda' of disabled people – such as high-tech prostheses, and also wheelchairs and crutches – can have a dual function in contemporary visual work. They not only act as semiotic markers of difference, but also as seductive performance invitations into a different form of embodiment.

In this chapter, I show how the aesthetics of presentation can engender a useful ambivalence towards the addenda of disabled people, undermining stereotypes of tragedy and negativity. I argue that the origin of this potentially positive ambivalence towards disabled people's 'difference' can lie in the audience's address of visual work, and its ability to connote and cite physical reaction and kinaesthetic experience. Visual technology and performance modes of audience address can merge and create a sensual space of engagement in which the other's body becomes tentatively experiential. With my readings here, I offer the thesis that contemporary culture seems fascinated with structures of feeling for non-traditional embodiments, and that contemporary

169

visual technology allows us to indulge that fascination in ways that supplement or even replace live performance. I will address the nature of this sensual engagement with otherness by analysing relationships between the visual and the tactile, the surface and the visceral, and the visual tactility/tactile visuality discussed by Maurice Merleau-Ponty (1968).

Aimee Mullins's visual persona is my example of this fascination with non-traditional embodiment. She is a fashion model, athlete and activist who walks with leg prostheses. She embodies both the fascination of the 'other' – the exotic, strange and different – and, at the same time, her representations seem for me to hover on the edge of inviting me into her living experience. She does not remain 'other', but comes closer. In the following textual analyses of a TV commercial, I am stressing the ruptures and tensions between the image of Mullins and the phenomenologically accessible performance of that image. I follow this discussion with an analysis of the audience which addresses a group of participants employed in a community performance installation I facilitated for Digital Summer 2000 in Manchester, UK. In this technology-driven community arts project, we tried, with different means, to tap into kinaesthetic sharing as an avenue for contemporary political work.

Seeing physically: The different body

Interactions between the semiotic and the personal physical, kinaesthetic connection can be seen to be fundamentally a part of visual perception. For phenomenologist Maurice Merleau-Ponty in his last, unfinished work, visual representation is in a productive tension with a form of tactility, the physical extension of vision. This tactility of vision is an addendum to the visible space:

> We must habituate ourselves to think that every visible is cut out in the tangible, every tactile being in some manner promised to visibility and that there is encroachment, infringement, not only between the touched and the touching, but also between the tangible and the visible, which is encrusted in it, as, conversely, the tangible itself is not a nothingness of visibility, is not without visual existence. (Merleau-Ponty, 1968: 134)

This deep connection between tactility and visibility lies in the material sharing of the universes charted by the two senses, by the fact that our bodies are always in movement, and that moving eyes is a tactile act, even though that continuity of materiality is not something that normally impinges on our experiences of our visual sense.

> Every experience has always been given to me within the context of the movements of the look... every vision takes place somewhere in the tactile space. There is a double and crossed situating of the visible in the tangible and the tangible in the visible; the two maps are complete and yet do not merge into one. The two parts are total parts and yet are not superposable. (Merleau-Ponty, 1968: 134)

There are ethical implications of this continuity between the visual and the tactile (a point Laura Marks (1999) also marks out in her study of the uses of sense impressions in intercultural cinema). These senses are often associated with distance and nearness respectively. The dynamics of understanding oneself to be physically implicated in the act of looking mean that a note of sharing can impinge on the visual act:

> Since vision is a palpitation with the look, it must also be inscribed in the order of being that it discloses to us; he who looks must not himself be foreign to the world that he looks at. (Merleau-Ponty, 1968: 134)

This horizon of the visible in the tangible, the promise of visual closeness and the erotics of touch seem to me to be a crucial aspect of contemporary visual culture and its relationship to disability as glamorised difference.

Contemporary cultural images in the West seem to move away from a purely visual and distanced relationship to representation – instead, a lot of contemporary representations reference and play with the physical, the kinaesthetic, the 'structure of feeling' of specific kinds of embodiment. In particular, contemporary digital artwork is able to manipulate points of view in a manner not easily emulated by the physical camera apparatus (although already part of the cinematic apparatus, as Vivian Sobchack's writing on cinema and phenomenology reminds us (1992)). Digital viewpoints often attempt to viscerally recreate the experience of movement. Computer games become ever more adept at creating the gravitational experience of driving on a race circuit. In these instances, the tactility of viewing, that is the location of the eyes in relation to space, time and weight are manipulated.

Aimee Mullins: Seduction and other beauty

The eye seduced into spatiality and viscerality through the camera – this is the core appeal of the first example I want to discuss. In *Catwalk*, a television commercial for the Internet provider FreeServe, screened in the UK during April and May 2000, Aimee Mullins's body and her status as a carrier of female connoted metaphors and discourses are crucial to the functioning of the advertisement. At the same time, her exoticism is mediated through a highly charged, visceral audience address – an echo of embodiment reaches beyond the purely visual. With this, the advertisement uses Mullins in a way similar to Matthew Barney, when he introduces Mullins in the *Cremaster* series of art films (2003). In these art films, Mullins functions as a borderline figure between humanity and animality, and she is not only wearing cheetah-like leg prostheses, but also a full-body costume presenting her to us as an animal. The TV advertisement is more subtle in the way that human, cyborg and animal identities interweave.

In the *Catwalk* TV advertisement (2000), a child's voice leads us into an experience – 'What do I like about Aimee?'. A relationship of desire is set up as a quick montage of shots introduce the spectator to a high octane, youth-dominated and sexy version of the world of the fashion catwalk.

The beginning of the video sets up the fashion context through a montage of relatively stable shots of a head, made-up in wild colours, hair sticking out – a mannequin on a catwalk. But this stiffness of pose, still photo and lacquered paint are disrupted by the montage of these images together with shots that show Mullins in interaction. First, the head shots are intercut with her playing with the child, whose voice 'Let's see – we have a lot in common. She likes to run, like me . . .' links the first images together. In these images, Mullins is a young model – nothing connotes her difference, and the only 'addenda' that we see are false eyelashes, lots of colour, hairdresses and fashion extensions in the head shots. Mullins' artificial legs are introduced later in the commercial, cut into shots of a spontaneous party, which puts the model into a 'real' social situation, and shows her laughing amongst a number of people. They are steel springs which do not attempt to emulate a human lower leg. Metallic, alien, vaguely insect-like in their back – rather than front-bending – these legs open up a different semiotic register from the glue-covered doll's legs.

These spring legs come into closer view as a cut from a laughing face ambles over to the mechanics of the legs, and their sneaker-like, commonplace anchorage to Mullins' knees. The placing of this shot amongst hand-held, fast moving, 'atmosphere' shots emphasises the relative normality of the situation and Aimee's unusual legs to the stage workers and models involved. The shot travels from her legs up her body to her laughing face, and down again. Her hands move down her body and clasp the spring legs as she doubles over with laughter – a gesture kinaesthetically familiar to a person with flesh legs. These legs are not shot here as the exotic stills of Mullins' model stage persona, but as part of her social and everyday life.

From this party scene, having brought us from Mullins, the exotic model to Mullins the right-on girl, the spectator hears Mullins' voice, addressing us, with her face looking straight at us. The snatches addressed to us remain vague: '...the one cheetah just stared, just stared...', 'you gotta run with the cheetahs and the antelopes...' 'it's a total feeling of accomplishment' (*Catwalk* 2000). In the visuals that accompany these statements, the relatively naturalistic spatial stability of the catwalk/pre- and post-show parties locales are disrupted with a cheetah's head, turning from side to side, and shots of Mullins' running on the catwalk in an 'animal' outfit with whipping feathers, including close-ups of the spring legs. I want to argue that in this part of the commercial it is not the child's voice and presence, the familiar exoticism of model's make-up, or the party atmosphere that bind the spectator to the image, but a kinaesthetic, tactile engagement of visuality.

As the animal narrative begins, we see Mullins' torso, moving in a slow, steady step, which is taken up by the turning of the cheetah's head in the intercuts. The cheetah is tracking something with its eyes – and the movement of vision is captured in the image. The timing of the steps/swings/paces steps up slowly, as we see the metal legs lift up from the floor, and Mullins' feather headgear bouncing in the rhythm. This visual rhythm culminates in a sprint. In the climax of the commercial, and the implied climax of the couture show, Mullins runs down the catwalk, and the run is captured in close detail, bringing together the exoticism of the animal spirit and the technology of camera and high-tech spring legs. Narrative and format mesh together to draw the spectator into the embodied experience of Mullins – the film stresses her understanding of herself, and the freedom of her legs, and her fascination with speed. This message is the one aimed at by the advertisers – 'be free' becomes 'be freeserve' on the screen (*Catwalk* 2000).

The lush surfaces and textures of these images, and the speed of cutting invite a tactile engagement *in* the visual engagement – and therefore disrupts for me any placement of Aimee as purely 'other' (purely visual spectacle, purely construct, purely exotic, animalistic stranger). The animal imagery stands in counter-point to the glinting metal spring legs. At the same time, the strange is placed into familiarity through the sensual appeal of the fast sequence, aligning me as spectator to the 'freedom' (of freeserve) and to a (potential) echo of the speed of Aimee's cyborg body. The addenda of the visible, the tactile, puts under erasure the addenda of the body, the cyborgian supplements, and on a trajectory towards unified engagement with a lived experience, my own and, potentially, Aimee Mullins'.

Body spaces: A disability culture tech-performance

Moving towards the other body as a form of seduction goes beyond a spectator address that surrounds him or her with fairground mirrors of difference. Instead, this form of seduction aims at an (equally narcissistic, but less distanced) fantasy play of 'being different'. It is this form of engagement, and the tensions that pertain to it, that fellow collaborators and I explored in *Body Spaces* (2000).

This show, commissioned by and presented as part of Digital Summer 2000 in three site-specific installations in Manchester, UK, emerged from a residency with young disabled people. Its creation was fuelled by an interest in the relationship between bodies, addenda and space. Our shared architectural/environmental framework brings bodies, senses and space together in particular narratives. This framework creates a habit that creates and is created by normative bodies – bodies that fit its doorways, pavements, dimensions, transport systems. The framework is normalised, invisible in everyday movement for those whose bodies fit its criteria. But for different bodies, its blueprinting mechanism becomes highly visible/tangible. Bodies with permanent and temporary extensions, such as wheelchairs, crutches, walking sticks or baby buggies, find themselves in conflict with the culturally constructed framework of spatiality, and their presence make the framework visible as a construction.

In the residency, many of us were wheelchair-users and people using crutches. As an occasional wheelchair-user myself, and one who moved out of the city into the countryside due to this blueprinting, more tangible in dense human-made spaces, I was fascinated by the idea of re-colonising different urban environments. Together, we created environments that choreograph the spectator's physical experience, that send him or her on a trajectory towards difference, and that distance their spatial/visual/tactile experience from the normative.

In the workshops, we discussed these implications of architecture and normalisation, and decided to show up our normative processes in a playful, productive and pleasurable manner. We were not interested in a continuation of the normative process which validates some spatial embodiment but not others. We did not desire an approach aimed at inducing guilt or discomfort – we wanted to show our many different kinds of locomotions and their relation to our environment, but without value judgement. Instead, we attempted to undermine the conventional stories of restriction and tragedy that are attached to other bodies. I will elaborate on some of the multiple strategies that we used to foreground phenomenological experiences of environment, embodiment and communication. In them, many of the audience address strategies I described in the Mullins advertisement resurface in our play between spatial technologies of movement and media technologies.

Strategy 1: Photography

The distinction between visual apprehension and tactile, sculptural work is made redundant in the sensual universe mapped by Merleau-Ponty. But within our everyday world, vision is a privileged sense, a supposedly distanced, surveying, dominating approach to the world that disavows the viewing self and that abstracts experience. One of the strategies used in the installation attempted to engage this distancing connotation of visual engagement by corporealising vision, making viewpoint and embodied vision 'visible'.

In all three installation sites, we were surrounded by everyday objects – objects vital to the functioning of the spaces, but invisible to the everyday eye due to their utter, banal normality. In storytelling and drama sessions, we started to spin stories around these everyday objects – telephones in waiting areas, wastepaper baskets, no-smoking signs in theatres, disabled parking spaces and leaves in the car park. Through these stories, the participant's recurrent fantasies, preoccupations and wishes became tangible – doors led to rock-star recording studios, empty seats were haunted by ghosts.

Taking her cue from these newly invested everyday objects, the production's video artist, Sara Domville Maguire worked with the young people to capture the objects from their own perspective. The participants began to establish the ownership of their vision, seeing an object deliberately from their own (often below the level of the normalised) position in wheelchairs. These photos of weirdly looming telephones, new and unusual arrangements of colour and blocks, close-ups and angled vision became an important part of the final installation. They were hung or laid around the installation space, providing a form of pilgrimage, a route through the site, traces of physical journeys. They were displayed in such a manner that the 'neutral' vision position was not possible – you had to crane your neck, or stoop down to see the seductive, brilliant, colourful new arrangements of space. Together with the stories, strewn in the vicinity of the objects, these photos became evidence for the embodied vision onto the world, an image invested with a spatial self, an act of fantasy and muscular, tactile activity, a physical interaction with camera and space.

Strategy 2: Pathways

The route around the memory/play sites, the photos and stories was mapped onto the terrain. During the residency, we investigated the different forms that claims to space attending wheelchair-use takes the use of crutches and use of feet. Just as sliding on the ground creates a new spatial experience and familiarity with the floor and the terrain, so does the continuous, round motion of the wheelchair, which never (usually) leaves the ground, and remains in touch with it. Walking as a biped creates digital, on/off, impressions on the ground, and a rhythm to movement that is different to the rhythm of the hand propelling chair-wheels, or the finger controlling an electric wheelchair. Spatial orientation is different – an electric wheelchair can have a majestic, full, round turning circle, or the three-turn motion of a car turning, while turning on one's feet provides another experience in which different body parts are facing different directions. In the installation, we were interested in drawing the spectator's attention to these different techno-logies of moving in space – the different kinds of alignments, motions, sequences. Using tape and chalk, we mapped lines on the ground – similar to the lines that denote different pathways to different sections in many hospitals – an interesting medical connotation in relation to disability. One line, following a wheelchair's wheel, moves continu-ously. It provides a visual continuity, like a rail moving across the terrain. Its neighbour, sometimes crossing and weaving across the chair line, the 'feet' line is actually more unusual to the observing eye – small indi-vidual bits of tape or lines of chalk which climb onto the horizon. In the installation, these lines were measured and adjusted to the stride of a particular person – and the attempts of visitors to 'match' that stride – to insert themselves bodily into these traces were an interesting sight.

Different from the convention of visual art practice, the creators of this installation were always in the space, linking the difference hinted at in the artefacts to real people, in real chairs, colonising a space where

Figure 20 Installation Still, *Body Spaces*, Petra Kuppers, © 2000

they are not usually seen in these numbers. We watched the visitors, witnessing their interactions with the photos, the terrain traces, the sensor elements, and we intervened into their explorations when we felt it right to do so. Those who wished wheeled up to visitors, and invited them to 'walk the installation path' with them, displaying their pride in their creation, their poems, their visuals. This 'live element', working across lines of performance and display, stage and off-stage, proved a very interesting strategy and intervention.

Strategy 3: Video

Part of the installation were projections of a performance-video shot with the participants, projected onto different surfaces. These videos could be manipulated by interactive controls offered to the visitors. The projections provided the main visual draw – they were visible from farther away, and drew spectators towards the quieter elements of the installation. In one of the installations, the video was backprojected onto a plastic plane spanned onto the back of a truck in a car-park. As night and rain fell, passers-by witnessed a jewel-bright display reflecting off rain-drops and wet pavements (*Bodyspaces* 2000).

The video *Geometries* investigates the geometries of bodies, their boundaries, and the shapes and volumes afforded by permanent and temporary addenda (a wheelchair wheeling by itself on the ground, driven by a performer moving around it; crutches and fingers; a wheel touched and explored by hands; toes and the wheel-spokes; the metallic refection of a chair). Instead of focusing on narrative and the insertion of wheelchair-users into a social space (the performative action of the whole installation), the video focused on qualities of shape and line. The video detached the disability experience from its social, medical and oppressive moorings. At the same time, the simultaneous occupation of the given space by the video, the traces of our life and imagination, and our live presences did not allow the video to become merely an exploration of formal beauty and human movement.

The interfacing modes offered by the installation foregrounded vision and its texture, scope, shape and weight. Examples included 'walking the lines', activating curiosity and desire by seducing the visitor to look up or down, round corners, to the photos and videos and providing various interactive devices that actively linked space habitation to vision (and sound[1]). A radio mouse sown into a curtain or left on the seat of a comfortable sofa (different in different installation sites) allowed visitors to move the video along, pause and play (via a QuickTime movie and a Director script). You needed to move in order to see – in an extension of Merleau-Ponty's moving eyeballs, here hand-co-ordination became the ground for visibility. A radio mouse was completely unfamiliar to many visitors, and people tried to explore the ways in which communication was established, how impulses were 'bodily' transmitted. Many used this 'disconnected' mouse and its functions as a way to start up conversations with participants.

Body Spaces (2000) provided its visitors with a playground and a dense space, full of traces of living, and of living differently. Many passers-by took up the invitation, and engaged in play, readings, communications and discussions with the participants.[2] Visibility and kinaesthesia created stories – things to explore, or paths to take, or ways to become aware of one's own constitution in physical space vis-à-vis experiences of disability and other physicalities. I think of *Body Spaces* as an apparatus of difference, a machine that allowed for experiencing one's body differently, not too far removed from a mirror cabinet on a fairground. By placing this machine into the everyday, into the flow of urban life, it provided an obstacle, inviting negotiation. Some passed by, some stopped and looked from the outside, some entered the space. Some lingered and played with image and text, others started to talk with us. Some took up our invitation to add their own fantasies and stories to ours, displayed on sheets on the ground. Without totalising the space into one belonging to us, we hope we inserted both a visibility and a tangibility of difference into the flow of the Contact Theatre in Manchester, the car park outside the theatre, and the Manchester Royal Infirmary Outpatients Lounge.

As an example of contemporary performance work, *Body Spaces* is an investigation of identity politics between and across the image and the body. This crossing, so well explored in contemporary computer games and highly affective, physical, kinaesthetic computer-enabled advertising, became useful for our Olimpias' community arts practice. As the visceral vision of the computer game allows for new ways of finding political expression,[3] performance artists embrace input-devices like joysticks and addenda like VR helmets. The wheels of chairs, which put some of us into space in interesting ways, need to become unstuck from narratives and images of tragedy and loss. It is up to us to re-invent the visceral meaning of all addenda, and make them tools in the re-invention of our social spaces.

Conclusion

In the different visual and performance works discussed in this chapter, I traced the engagement of the physical and kinaesthetical in interaction with semiotic layers. Merleau-Ponty's words point the way to an understanding of visual engagement as a political, ethical act – what is seen and the seer are part of a life world, of an extended physical reality beyond the boundaries of an individual body. The examples discussed here can be seen in this light – referencing sharing through the connotations of embodied experience. Through opening up these registers of continuity, the works can also be seen to play on the boundaries of the flesh/artifice/cyborg, both in their audience address as artefacts echoing physical experience, and as depictions of human 'addendas' which become part of different forms of embodiment.

In the contemporary representational field, the political ambivalences of semiotic registers do not vanish, but take on an additional charge as bodied experience seems to become the target of ever more involved technical extensions of perception and visuality. Blindness and its kinaesthetic realm become the point of excitement in films such as *Daredevil* (2004), and while 'real-life' visual impairment is not an issue here, only super-crip super-hero ability, the search for new forms of visual excitement and stimulation, leads to interesting avenues for physicality and difference. Hopefully, exciting bodies can emerge out of popular culture's desire for new forms of embodiment, and new opportunities can open up for disabled people. Visual excitement and ethical care for someone who is different are two different issues – but as our shared culture becomes more and more curious, maybe we can also grow more respectful.

Notes

1. Sound was also used extensively in the installation. The sound artists Sam Richards (an artist with an invisible disability) and Sarah Frances created a piece of sound art reflection 'inner sound', sound of living, the soundtrack for the video. Soundbytes from this track were also linked to motion sensor pads on the ground, part of the installation track, that allowed visitors to create their own soundscapes through movement.
2. Visitor comments can be found at http://members.aol.com/aerfen/body-spaces.htm.
3. For instance, in the embodied computer game of Blast Theory's *Desert Storm*, also shown as part of the Digital Summer 2000 festival. In this large-scale performance environment, spectators find themselves actors in a desert war situation.

References

Marks, Laura U. (1999). *The Skin of the Film: Intercultural Cinema, Embodiment, and the Senses.* Durham, NC: Duke University Press.
Merleau-Ponty, Maurice. (1968). *The Visible and the Invisible.* Evanston: Northwestern University Press.
Sobchack, Vivian. (1992). *The Address of the Eye: A Phenomenology of Film Experience.* Princeton: Princeton University Press.

Filmography

Catwalk. (2000). Advertisement produced by M&C Saatchi, for FreeServe, UK.
Cremaster 3. (2003). Dir. Matthew Barney, USA.
Daredevil. (2004). Dir. Mark Steven Johnson, USA.

14
Technology as a Bridge to Audience Participation?

Christie Carson

Online technology has made it possible for audience relationships to be reinvented and reputations to be redrawn across the theatrical spectrum. But is this what has actually taken place in the online world or are traditional priorities and prejudices slowly being re-established? Digital technology and the Internet allow for a democratisation of the producer/audience member relationship, and increasingly, it is possible to think of a two-way form of communication which extends beyond polite applause within the theatre building. Theatre companies can, in theory, bring the audience into the creative process of theatre-making and extend exponentially the reach of their activities. The smallest of theatre companies can now have an international presence but is the theatre world taking proper advantage of the opportunities that are available?

In order to explore this question in relation to several of the larger theatres in Britain, I will look at three examples of digital projects that are aimed at involving younger audiences through online education programmes. I will compare the *Stagework* project, which involves the National Theatre, the Bristol Old Vic Theatre and the Birmingham Repertory Theatre, the *Exploring Shakespeare* project produced by the Royal Shakespeare Company, and the 'Adopt an Actor' Scheme run by Shakespeare's Globe Theatre. In each case, I will consider the extent to which a new kind of audience involvement is created and the implications of these new levels of access. Above all, I will look at the ways in which these theatres are redefining their relationships with their audiences through digital technology and, as a result, redefining their public images. I will consider how effective the approach taken might be in developing a new generation of theatregoers and in creating an engaged debate about the role of the theatre in the twenty-first century.

Through these examples, I will illustrate what I see as a marked difference between the approach taken to audience involvement by Shakespeare's Globe Theatre, which receives no government sponsorship, and the approach taken by the subsidised theatres. I would argue that while Shakespeare's Globe Theatre has taken a spontaneous and inventive approach to involving additional audiences through their Education Department, the National Theatre, the Bristol Old Vic, the Birmingham Repertory Theatre and the Royal Shakespeare Company have approached the new technology in a much more traditional way. I would suggest that the relationships created through the Globe Theatre programme present a more fundamental change to theatrical audience interaction by developing a two-way form of communication directly with students.

The funded institutional theatres, on the other hand, are working ever more closely with the governmental structures for education, packaging an experience of theatrical creation which fits into a highly prescribed format. As a result, while an exciting new kind of archive that involves theatre practitioners is being created through the two projects described, the results are descriptive of past activity rather than engaging new audiences in current and developing activity. I suggest that the subsidised theatres are, through these projects, taking on the role of the library, the archive and the curriculum development company and in so doing reinforce a one-way form of passive communication with their audiences. Shakespeare's Globe Theatre, on the other hand, draws its audience into an ongoing process in a participatory way, involving the audience in the Theatre's creative project.

Shakespeare's Globe Theatre has been severely criticised for its commercial and tourist-oriented approach to Shakespeare, yet I would argue that the approach taken in the area of technology helps to point out exactly how well this theatre has acted as a constant reminder to the mainstream theatres of their own internal prejudices. While I do not defend all of the artistic decisions made by the Globe Theatre's creative team, I suggest that the presence of this theatre on the landscape has forced considerable changes on the other funded theatres in terms of repertoire, ticket pricing and audience involvement in a project that is larger than the individual performance.

This shift of emphasis towards a commitment to a particular and subjective approach to theatrical production undermines the universalist claims of excellence put forwards by the National Theatre and the Royal Shakespeare Company in particular. I must concede that the position of the subsidised theatres is very difficult in that they are forced to try to accommodate their traditional audience base, on the one hand, and notions of excellence placed upon them through governmental policy, on the other. The projects I will describe tread a careful line which boasts the availability of exciting new resources for teaching and research delivered in a package that caters to traditional approaches to theatre and to education. Looking at these projects, in contrast to the work at Shakespeare's Globe Theatre, will reveal the underlying pressures that are restricting these theatres from developing truly innovative new ways of talking to audiences using technology that could reinvigorate the theatre as a centre of public debate in the twenty-first century.

Three examples of digital audience interaction

Stagework – National Theatre, Bristol Old Vic, Birmingham Repertory Theatre

My first example is a complex broadband project entitled *Stagework* (http://www.stagework.org.uk/), which was commissioned by Culture Online (part of the Department of Culture, Media and Sport), and was created for a schools audience by theatre companies working with professional web developers. This project documents the work of three important theatres. Four productions, *Henry V* and *His Dark Materials* at the National Theatre, *Beasts and Beauties* at the Bristol Old Vic and *The Crucible* at the Birmingham Repertory Theatre are represented through extensive audio-visual archives.

The project involves interviews with actors, directors and designers, as well as members of the technical teams. In some cases, extracts from rehearsal and performance are also included. The information in the archive is organised through a series of time-lines for each production that map out Performance, Creative, Technical and Production Administration schedules. Each production also includes an introduction, a plot synopsis and a rehearsal diary. This raw material is then arranged thematically on subjects such as 'nationality and race' and 'images of war' for *Henry V* and 'staging miracles' and 'fable to stage' for *Beasts and Beauties*. The theatre building is also used as a navigational device, allowing users to get a sense of the location and the roles of the people involved. Finally, lessons plans are provided for teachers and information is offered about how these materials can be used at various key stages. This section of the site also points users to archives of live events such as an interview of the National Theatre's Artistic Director, Nicholas Hytner and a round table discussion involving several members of the cast of *The Crucible*.

This project presents an exciting array of materials that can be used in a variety of ways by a wide range of users. The flexibility of the approach offers users the ability to interact with the material in a non-linear way. The full capabilities of this new medium for archiving and studying performance are fulfilled to a large degree. The involvement of a wide range of creative practitioners is documented and the user is given access to discussions about the process by both individuals and groups reflecting on the rehearsal period and the interpretive process. Undoubtedly this resource provides unparalleled access to materials that describe the creative process of theatre-making. Students are given the opportunity to hear the views of a wide range of creative practitioners and can plot their own paths through the materials offered.

This project, however, extends the audience model of a passive individual in a darkened theatre rather than using the full live capabilities of the technology. The drawback of this approach is that it creates for the user the illusion of participation rather than providing real interaction. The excitement of the theatrical event, its liveness and unpredictability, is replaced by televisual staticness. Essentially what this project provides its audience is the raw material for a documentary. The audience member is given the opportunity to act as a television editor, creating a montage of already existing responses to questions posed by other people. The interaction is limited to listening, or not listening, to the previously prepared responses and rearranging those responses to suit one's own interests.

This approach, I suggest, raises worrying questions about the purpose of government-sponsored theatre when married to a government-directed curriculum. While I applaud the increased access to limited events like the National Theatre's Platform series, I question the approach taken in many of the interviews conducted with individual practitioners. In almost every case there is a rather odd vocational aspect to the questions which I find, not just confusing, but rather disturbing. The Education Director of the Birmingham Repertory Theatre, for example, is asked to detail his employment and education history. This information is then provided under headings such as 'What is an Education Director' and 'I want to be an Education Director'. This approach is repeated across the site setting up a strange sense that one aim of the project is to provide a theatrical recruitment service.

Equally worrying, from my point of view, are the lesson plans provided for teachers. While a lesson plan exists for each of the productions to support teachers preparing lessons in English and Drama, two lesson plans are also provided to support teachers preparing lessons in Citizenship and one for teachers of Religious Education. The underlying implication here seems to be that government-funded theatre provides lessons in citizenship and religious direction. While I would not question the ability of theatre to address these issues, I do question the appropriateness of making this connection in a schools-oriented project that uses particular productions from particular theatres in such a directed way.

Therefore while the materials created and the opportunities provided in this ground-breaking project are unprecedented, the articulation of some questionable principles within the structure of this project makes it far from ideal. The coming together of government funding, educational policy and artistic creativity places this project in uncomfortable territory. The question that arises is does this project articulate the beliefs held within the theatre community, that their work is designed to build better citizens and create new workers, or is it the result of creating a project within the strict structures of educational policy. This example, while boasting sophisticated technical prowess, lacks directness of purpose in terms of the relationship set up with the audience. The relationship that is established with the user is that of a voyeur not of a participant and the metaphor that is being invoked is passive and televisual.

This project model extends the work of these theatres and gives access to their working practices but in a way that is entirely controlled both in content and in form by the project's creators. The audience can have no impact on the process of theatre-making since the productions documented have long since left the stage. The users of this archive can only have access to that which was considered fit to protect for posterity. If giving students a sense of what is important and exciting about the theatrical event is the purpose of this project, I suggest it is misleading in that it takes away the two most important aspects of that event, liveness and real audience interaction.

Exploring Shakespeare: Hamlet and Macbeth – The Royal Shakespeare Company

The second project I will examine is the recently launched *Exploring Shakespeare: Hamlet and Macbeth* site developed by the Royal Shakespeare Company (http://www.rsc.org.uk/learning/hamletandmacbeth/). This multimedia site, like the *Stagework* project, provides an extensive audio-visual archive of two RSC productions, *Hamlet* directed by Artistic

Director Michael Boyd and *Macbeth* directed by Associate Director Dominic Cooke. Like the *Stagework* site interviews with actors, designers and directors are provided alongside video clips of performance and rehearsal. Again the material is not only provided in its raw form but also gathered together in a structured way to look at particular themes, 'Power and Control', 'Murder and Consequences' and 'The Tragic Hero' for *Macbeth* and 'Sex and Violence', 'Secrets and Lies' and 'Revenge' for *Hamlet*.

Maria Evans, RSC Director of Learning, says in the press release issued at the launch of the project:

> We are really excited about this project, and have already had some very good feedback from teachers who got a sneak preview of the site at our stand at this year's Education Show in Birmingham. We've aimed *Hamlet* at Key Stage 5 (16 and over) and *Macbeth* at Key Stage 3 and 4 (13 to 16 year olds). However, it is open to anyone who has access to Broadband. (Evans in RSC, 2005c)

This project also includes a section for teachers who are provided with a range of activities to use in class. The guidance for teachers looking for material on *Hamlet* is as follows:

> Activities are designed to help students experience the play in their own terms whilst supporting their ability to use the language of Shakespeare. Emphasis is on the interpretative choices made by a production in order to encourage students to have confidence in making their own creative judgements.[1]

The activities included, such as 'Writing a Review', 'Working on Design', 'Practical Scene Study' and 'Ways into Character', to name just a few, are provided to teachers as downloadable Word files. In fact this area of the site appears to make more widely accessible the teachers packs that have been a feature of the RSC's Education Department for some time.

There is real insight and innovation in this site but it appears to replicate the authoritative static nature of the *Stagework* project in a number of ways. Again I would suggest that the form in which the materials are made available says a great deal about the ethos of the theatre company and the image it would like to project. In this case, the Stavros S Niarchos Foundation has supported the project, leaving the RSC free to follow its own agenda rather than a proscribed educational pathway.

What emerges from this project are new means of archiving material and making it available to users. What also emerges is a new way of using a performance text as a navigational device, providing several routes to the materials on offer. While all of these innovations are welcome, I suggest the emphasis on the text is quite telling.

On arrival at the site, the user must choose first a play to look at and then a thematic area. Once the user enters one of the thematic areas a video screen appears and the video of either a rehearsal or performance of a particular scene in the play commences. As the scene proceeds, questions pop up on the screen which indicate with an arrow to the right that there is more information on this topic. Clicking on the question brings up a new video box in a window that offers an array of answers to the question from the cast and production team. The progress of the original scene stops for the duration of the parenthetical journey chosen by the user and then recommences when the extra window is closed. This clever device provides the opportunity to get an insight into the interpretative choices of the cast as well as the directorial and design choices that contributed to the final performance. Access to rehearsal room versions of the scenes in some cases and final performances of the scenes in others places an emphasis on the developmental creative process of theatre-making.

So performance is at the centre of this project but it is performance as a means of interpreting the text, rather than as a meaning making form of its own. This emphasis on the textual interpretation of the plays emerges through both the project's form and the content. An example of this in terms of form is the fact that it is possible to turn on and off captions that present the text of the play while it is spoken. This is also the case for the interviews of the individual participants. This is an obvious requirement for the purposes of accessibility both for the hearing impaired and for those users who do not possess speakers; however, I suggest there is a further meaning to this.

Each scene presented is divided into four parts and each part is identified by the first line of the text spoken in that section. Thus the text provides a navigational tool within the project and also serves to connect the online resource with the printed text outside the project. The text, then, can be seen as the organising principle as well as a functional element of the project's form. In terms of content, the centrality of the text and its interpretative context is further emphasised in the interviews. Michael Boyd, the RSC Artistic Director and director of *Hamlet*, discusses at length the importance of the differences between the Quarto and the Folio texts of this play. One of the actors in this production describes his passionate conviction that actors can find in the Folio text instructive material that comes directly from Shakespeare. These two interviews reveal an approach to the plays which, I suggest, says rather a lot about the RSC and its working ethos. What is made clear is the belief that the RSC actors and directors are part of a continuous interpretative theatrical tradition that stretches directly back to the Bard himself.

So while the *Stagework* project provides the students with an approach that is televisual, the *Exploring Shakespeare* project presents an approach that is far more influence by the publishing history of the texts and the performance history of the plays. In this case there are no attempts to recruit new theatre practitioners from the audience. Rather there is a reverential approach both to the plays and to their interpretation. Again, like the *Stagework* project, the questions posed are not at the discretion of the users and the productions documented have left the stage some time ago. Here also is the presentation of an archive and a teaching resource rather than an interactive event. While the footage that is made available again exceeds dramatically what has been available in the past, there is still very much a sense in which the audience member remains passive and at a distance. The communication is entirely one way and the user can have no impact on the activity of the RSC. This very contained and controlled approach is both safe and secure. The RSC seem to see no need to involve the audience through this technology rather it sees the service it presents as extending access to work that already exists. This approach seems to lack either an understanding of the power of the new technology or a distrust of its democratising principles.

While this project is not funded by the government as part of it educational policy the teacher's resources are very clearly aimed at particular areas of the curriculum. Key stages 3–5 are addressed with particular assignments and the project seems to see no contradiction in providing quite prescriptive activities. The increasing push towards the standardisation of the curriculum is something that the government is now moving into higher education. The development of Higher Education Funding Council for England's (HEFCE) Centres for Excellence in Teaching and Learning (CETL) shows a strong steer by the government to again celebrate the work of exemplary approaches rather than encouraging diversity of approach. The RSC's partnership with Warwick University in one such Centre seems to illustrate how this theatre company has not been immune to the pull of central funding directed at the standardisation of education (RSC, 2005a). In fact, given that the Company has stated clearly that a concrete outcome of the 2006 Festival of Shakespeare's Complete Works will be 'A report . . . [that] will make a series of recommendations to the Government and key policy makers on the future teaching of Shakespeare' (RSC, 2005b) the RSC's position as authority in this matter seems undisputed.

'Adopt an Actor' Scheme – Shakespeare's Globe Theatre

The final example of an online education project I will examine differs distinctly from those I have already described. For some time, Shakespeare's Globe Theatre has run its 'Adopt an Actor' Scheme through its Education Department.[2] The scheme involves student groups being assigned to a particular actor for the duration of the rehearsal process. These groups may be local and the students may watch the live performance at the culmination of the process, but equally the group may be at a distance, even overseas, resulting in the online communication being the only real interaction with the theatre. By giving students the opportunity to communicate directly with an actor through video conferencing establishes a direct and real rapport that is ongoing. The students are able to follow 'an actor's experience as they create a role from the first day of the rehearsal to the final performance in the Globe' (Globe, 2003: 20).

In this Scheme, the actor reports to the students through a series of rehearsal notes on the progress of rehearsals and the students are asked to respond directly with suggestions as well as comments. As the website describes:

> GlobeLink's bespoke videoconferencing programme provides students of all ages with virtual access to the Globe Theatre and the discoveries being made there by Globe Education and the Globe Theatre Company. Students around the world are able to gain insights into the playing conditions of the Elizabethan playhouse and how they affect our understanding of Shakespeare's plays through online conferences with actors, directors, designers and other theatre professionals.[3]

During the 2005 season over 2000 students worldwide will be involved in the 'Adopt an Actor' initiative. In the past this programme has facilitated the incorporations of suggestions by the students into the rehearsal process. For example one actor asked for ideas of a hobby that the actor's character could pursue. One suggestion, collecting stamps, was incorporated in the initial rehearsals but was eventually discarded once the rehearsals moved on stage, the stamps being too small for the audience to see. Despite this outcome this process created not just a sense of involvement in the production, it gave real power to the students to see their ideas enacted.

This scheme, which is run on a shoestring through the voluntary work of the actors involved, makes creative use of digital technology to engage students in the theatrical process. The Globe's commitment to 'practice-led resources for the study of Shakespeare' (Globe, 2003: 20) is amply demonstrated through the active involvement in the creative process that is facilitated. This programme does not try to address the production as a whole, although other areas of Globe Education do, rather it makes a virtue out of the fact that the students are given a personal and extremely subjective approach to the play. To see the play through the eyes of just one character, as the production evolves, creates for the students a personal journey that they will likely remember forever. The direct and personal nature of the communication set up mirrors the communication established in the open-air daylight filled theatre. As Mark Rylance, Shakespeare's Globe's Artistic Director, says of the theatre 'here you are aware of how much Shakespeare kept the audience in mind and what a difference there is when an actor is able to do that too' (Rylance in Shenton, 2002: 55). The overlap between the ethos of the theatre and the education programme is obvious in this activity that so directly links the students with the creative process.

This is a link which carries over to all of the Globe Theatre's activities. In fact no other theatre in Britain is comparable in dividing its time between its educational activities and its theatrical activities, splitting the year and its staff equally between these two joint aims. As a result, a great many of the Theatre's actors and musicians are also Globe Education Practitioners. To extend the activity of the theatre to an international audience through digital technology is a logical extension of this theatre's current working practices and theatrical ethos. Shakespeare's Globe Theatre is clearly and passionately committed to a particular approach to Shakespearean performance and to education. Patrick Spottiswoode, Director of Globe Education says:

> At our playhouse, students and teachers meet and work with people who revive words and help them play. Actors, directors, musicians, voice and movement coaches, designers and fight directors, who rouse words into lively action share their craft and passion with over 50,000 students and teachers every year. (Spottiswoode, 2003: 1)

This commitment to a way of working that involves students and teachers at all levels in every aspect of the work of the theatre creates an engagement with the theatrical event that is embodied in all of the work of Shakespeare's Globe Theatre.

There are some who would consider this approach one of evangelical indoctrination. I suggest rather it is an exposure to a community of like-minded individuals who work together in a collaborative and engaging way. Exposure to that community and to that working ethos provides students with an example of commitment that is not matched in many other areas of their lives. The devotion of the actors, who volunteer their time to speak about their work directly to students across the world, says something quite interesting about this commercially funded theatre. With only the most rudimentary of equipment Shakespeare's Globe Theatre has created a programme that has the potential to create a truly transformative experience for its participants. Learning about the creative process through a particular character in a particular play inevitably is a concentrated, case study approach to the work of the Bard. But it is an approach that has the potential to instil in the student participants a very clear sense of the purpose and relevance of these ancient words through a personal and private dialogue set up between audience member and actor. The participatory, non-hierarchical nature of the communication established in this model shows the real power and simplicity of the technology in its purest form.

Conclusions

The three projects described illustrate very interesting patterns in funding and in the development of educational materials, as well as the attitudes of the theatre companies involved towards the development of new audiences. While each of these projects is designed specifically for a schools audience, only the Globe Theatre's 'Adopt an Actor' programme speaks directly to that audience in an open-ended way in real time. The *Stagework* project and the *Exploring Shakespeare* project, by contrast, present educational materials that document in detail the creative process of individual productions. This material is freely available to the public as well as a student audience. While this is a worthy approach there is no denying the fact that since these projects present unprecedented access to the work of these companies they have the potential to develop a vision of that work in the popular imagination. The fact that the subsidised theatres have been drawn, at least to some extent, into the government's drive towards standardisation of educational provision through funding initiatives, I suggest, raises some awkward questions about the role of these theatres in our society.

I must acknowledge that both the National Theatre and the RSC have introduced a range of live events that draw a general audience into a variety of activities and debates; however, these events continue to centre on the gathering of individuals at their respective theatres. The National Theatre's summer programme of riverside events entitled 'Watch This Space' includes an enormous array of free outdoor performances, roundtable discussions and a late lounge dance venue. All of this activity, while exceptionally inclusive, relies on the audience member's ability to come to London. Similarly, the RSC has hosted a number of events that have been designed to involve the public, including an enormous celebration of Shakespeare's birthday; however, these events assume a presence in Stratford-upon-Avon. The extraordinary success of the international Live 8 event in combining live, television and online audiences in real time must stand as an example for these theatres of the way this technology can be used to involve a wider group of people in a more direct way.

In the web projects described the work of the subsidised theatres has been showcased but it is work that has already been presented in other forms in other places. The full capabilities of the Internet to involve an audience who are geographically spread across the UK in the ongoing work of these theatres have not been exploited. The subsidised theatres are being asked to present a model of exemplary British theatre craft rather than providing a centre of social and cultural engagement. The Globe's online work shows how straightforward it can be to engage an audience in the creative process and also in a direct discussion about the contemporary relevance of this theatre's work. The subsidised theatres, however, continue to pursue their historical position as preservers of cultural quality. In doing this I suggest that tax funding is being used more to instil conservative approaches to theatre than to engage a wide and varied local audience.

The online environment provides an opportunity for unprecedented outreach to a British audience that is not currently engaged with these theatres, as well as a student audience. Unfortunately, at present, these theatres have chosen to simply extend their current educational offerings to a wider schools audience. I suggest that, rather than solidify their position as exemplars in their fields, this move opens these theatres up to the criticism of cultural imperialism. Like many of us in education the large institutional theatres are being drawn in this direction through funding strategies. But I argue that the government's programme of social engineering should not be mingled with the creativity of the country's finest theatre artists. Despite the ingenuity and resources that have been invested in these projects, their close association with a restrictive curriculum results in a public display of the prejudices held, if not by these theatres, then at least by their funders. Given the enormous opportunities that this technology holds for creating a bridge to a new audience, as well as a discussion of the role of theatre in our society in the twenty-first century, this seems a terrible shame.

Notes

1. From the 'For Teachers' page on *Hamlet* at http://www.rsc.org.uk/learning/hamletandmacbeth/teachers/forteachershamlet.
2. Further information about Shakespeare's Globe Education activities can be found at http://www.shakespeares-globe.org/navigation/frameset.htm.
3. Further information about the 'Adopt an Actor' programme at http://www.shakespeares-globe.org/navigation/frameset.htm under Globelink.

References

Exploring Shakespeare: Hamlet and Macbeth (http://www.rsc.org.uk/learning/hamletandmacbeth/).

Globe (2003) *The Globe Education Brochure*. January–December.

RSC (2005a) '£4.5 million Performance Partnership for Royal Shakespeare Company and University of Warwick', RSC Press Release, 27 January.

RSC (2005b) 'RSC host first every Festival of Shakespeare's Complete Works in Stratford-upon-Avon', RSC Press Release, 11 July.

RSC (2005c) 'RSC Launches New Interactive Multimedia Site', RSC Press Release, 13 June.

Shenton, M. (2002) 'Smart Ass: Classical Theatre Superstar Mark Rylance Tells Mark Shenton How Shakespeare's Globe Lets Him be Intimate with Audiences', *Sunday Express*, 4 August: 54–55.

Spottiswoode, P. (2003) 'Word Play', *The Globe Education Brochure*. January–December.

Stagework (http://www.stagework.org.uk/).

Afterword: Is There Life after *Liveness?*

Philip Auslander

My inquiry into the cultural status of performance and media in *Liveness: Perform-ance in a Mediatized Culture* (Auslander, 1999) departed from two major premises. The first, borrowed from Walter Benjamin (1969: 222), was that 'human sense perception...is determined not only by nature but by historical circumstances as well.' It seemed to me important to insist that sense perception is not simply a biological given: rather, how we perceive and what we expect from the objects of our perception are culturally and historically influenced. That influence is reified in media, which are simultaneously cause and effect of a given historical moment's social formations and technological capabilities, a process that Benjamin tracks from painting to photography to cinema.

The second crucial premise was that media are not equal. As Alain Busson points out, each new medium struggles for cultural territory with existing media, a struggle whose outcome is that some media are dominant within a given culture at a given moment, while others are dominated (quoted in Pavis, 1992: 113). Other theorists have adopted similarly agonistic concepts of the interrelations among media; Jay Bolter and Richard Grusin's (1996) idea of 'remediation' is an example. My favorite formulation remains Marshall McLuhan's (1964: 158): 'A new medium is never an addition to an old one, nor does it leave the old one in peace. It never ceases to oppress the older media until it finds new shapes and positions for them.'

This way of thinking suggested to me a model of what I called 'cultural economy', a schema in which different media enjoy different degrees of cultural presence, power and prestige at different points in time. As the agonistic models I have summarised briefly here indicate, it is technological change, the appearance of new media that produces shifts in this economy in terms of which media are dominant and which dominated.

Taking these two premises together, I proposed that audience perception was likely to be most influenced by the dominant media of the time and that spec-tators would bring expectations based on that influence to bear on their experi-ences of non-dominant media. In my narrative, television (in its extended form as 'the televisual') assumed the role of dominant medium, while all forms of live performance were relegated to the position of dominated media. As Herbert Blau indicates, the current audience's 'odd anonymous needs gather around the luminous spot of the video tube, the hegemony of mass media...' (2003: 275). I believed then, and continue to believe, that interactions among media, and between live and mediatized forms, need to be understood in relation to a concept of dominant media.

Historically, the discourse around multimedia (particularly the discourse around the idea of combining theatre with film Steve Dixon discusses here) has not taken this tack. Rather, there has been a tendency to assume that all media are of equal cultural impact and will therefore command equal audience attention when combined in various ways. Different media may be deployed to mean different things: while Allardyce Nicoll aligned the stage with illusion and the screen with truth, Robert Edmond Jones felt that the stage conveyed the characters' 'outer life', the screen could convey their 'inner life'. (Allow me to note in passing my bemusement at the idea that the concrete presence of human beings is understood in both cases to represent mere appearance: for something more genuine than that, Nicoll and Jones both felt we need the screen.)

The notion that, working together, stage and screen can convey a fuller sense of what it is to be human than either can alone is premised on the assumption of their working together as complementary equals. The possibility that audience perception may inevitably be drawn to a screen even when there are human beings also present, for instance, is not usually considered as part of the equation. I must say that I remain skeptical of discussions of work based on the interaction of live and mediatized performance that do not factor in such considerations. As Robert Wechsler points out in his contribution to this collection, one reason why technical media compel attention is quite simply the 'how'd-they-do-it' factor. The audience's inevitable curiosity about how technical effects are achieved makes them a centre of attention, let alone the fact that such effects may reiterate the audience's experience of dominant media forms. Although some performance makers seek transparency in their uses of technology or to demystify the apparatus, it is not at all clear that such tactics actually derail an audience's fascination with technological spectacle and novelty. As practitioners and critics, we ignore such considerations at our own peril.

I do not wish to oversimplify, however. Though I think it is crucial to keep in mind that different media or modes of expression do not interact on a level cultural playing field, I am not suggesting that one cannot modulate the relationship among different elements at different moments of a performance. Clearly, there are ways of asserting the presence of a human body over that of a projection, for instance, or vice-versa. That does not change the fact that the performance occurs in a cultural context in which the projection is more closely related to the dominant media than is the live body, a fact that has implications for how the audience perceives the whole performance. But the cultural terrain is uneven in other ways, too. Turning from the question of dominance to that of prestige, for instance, one may find a somewhat different story. As Martin Barker (2003) suggests, even though the theatre has, in my terms, much lower cultural presence and power than, say, cinema or the Internet, it may enjoy greater prestige because it continues to be perceived as a high art form requiring specific educational and cultural capital to appreciate. Even though most people now would prefer to watch television or play a computer game than go to the theatre, they may still accord the theatre greater prestige. Video on stage or in an installation may thus become 'art', while video on your television set remains 'entertainment'.

The current version of this jockeying for position within cultural economy, unfolding as I write this, indeed involves computer games, which either are or are about to become a dominant medium with respect to capitalisation, cultural presence, and power. Certainly, they have already begun imposing pressure on other media. Hollywood film director Robert Zemeckis, for example, notes 'In the '80s, cinema became influenced by the pace and style of television commercials. And in the '90s, it was the pace and style of the music video. And I think the next decades are going to be influenced greatly by the digital world of gaming' (quoted in Breznican, 2004). Johannes Birringer proposes in his essay here that gaming may serve as a useful model for more experimental art practices as well, since the level of engagement and the sense of community produced by multi-player games seems to far surpass that of earlier forms of interactive art. (A vein of healthy skepticism concerning the value of interactivity as an end in itself and the ways it has been incorporated into art practices runs through this collection.) Increasingly, participants will bring the perceptual tendencies and expectations developed through their experience of gaming to other cultural forms. Nevertheless, computer games have not achieved cultural prestige: they are still firmly ensconced in the category of 'entertainment' even as other practices based on gaming and employing game engines ascend to the category of 'art'.

Zemeckis's comment also suggests that it is time to posit the digital, rather than the televisual, as the current cultural dominant. At the time of writing *Liveness* (based on work I had been doing since the early 1990s), I was reluctant to make that move. But it now seems clear that we experience the televisual or, at least, television, largely through its digital remediations (on the Internet or DVD, for instance). It once seemed that television could absorb any cultural discourse and turn it into itself; now, that capability characterizes the digital, which has absorbed the televisual among other discourses. Sue Broadhurst gives a hint of what human perception conditioned by the dominance of digital media may be like when she refers to ' "indifferent differentiation".... A [way of] thinking that makes little distinction between the referent and meaning or for that matter between "reality" and representation' (see Chapter 11). Two recent literary scandals come to mind in this connection: author James Frey was found to have fabricated parts of his memoir, while fiction writer J.T. LeRoy was found to be a work of fiction himself. Are the ensuing discussions of whether or not this matters to the value of their respective works early symptoms of the indifference to such distinctions that Broadhurst identifies as central to the digital episteme? Does the potential breakdown of the real and the fictional as distinct categories presage the way writing will be shaped and positioned in a cultural economy dominated by digital media?

Jean Baudrillard's (1990: 155–156) description of the digital as the undoing of both dualities and polarities in favor of norms and models anticipates this development. One of the arguments that those who support Frey and LeRoy offer is that as long as their writings conform to the model of a kind of literature in which the protagonist wallows in, but eventually overcomes, abjection, it serves its inspirational purpose whether or not it is factual. Since *Liveness*, I have come to think of that concept more and more as a moving target that, unmoored from ontology, takes on different meanings at different times. In its classic formulation, liveness implies both physical and temporal co-presence of performer and audience; this corresponds to Baudrillard's order of duality. With the advent of broadcasting, the concept of the live referred only to a temporal relationship (as in a live broadcast), which corresponds to Baudrillard's order of polarity (the polarity of live vs recorded when both were conveyed by the same media). Now, when a website becomes available for interaction online, we say it has 'gone live' regardless of its content. This suggests that the defining quality of the live at this point is feedback – we accept any situation in which we receive a signal in response to one we have sent out as a live interaction. We have, as Baudrillard suggests, moved decisively from a cultural order characterized by 'relations' among things to the digital order characterized by 'connections' between things.

One implication is that whereas liveness once connoted a-liveness, that is no longer necessarily the case, a situation nimbly dramatized by the work of the Tissue Culture & Art Project (TC&A) described here by Oron Catts and Ionat Zurr. The Semi-Living entities created by are live performers, yet they both are and are not alive in the usual sense. No longer just a duality, alive and non-living become points on a smooth continuum of liveness with (perhaps) human performers (live and alive) at one pole and performing technologies (live but not alive) at the other. That TC&A's Semi-Livings would have to be placed at some undefined point in between problematises the very polarity along which the continuum is mapped.

I note with interest that Catts, Zurr and Broadhurst are the only contributors to this rich and nuanced collection to discuss work involving non-human performers (though both essays make it very clear that the non-human is inevitably imbricated with the human). Most of the essays are decidedly humanistic (or post-posthumanistic, if you will) in emphasis, reassuring us that technology cannot take the place of human presence at the heart of performance, that it is best used to extend the capabilities of human performers, to express humanistic themes more fully, and to allow performance to explore or evoke responses from realms of human physical and psychological experience not directly accessible otherwise (including the physicality of the disabled, which is not experientially available to the able-bodied under normal circumstances, as Petra Kuppers indicates). Perhaps, then, our anthropocentrism is the territory we are not willing to cede to the dominance of the digital, at least not now. Or not yet.

References

Auslander, Philip. (1999). *Liveness: Peformance in a Mediatized Culture*. London, New York: Routledge.

Barker, Martin. (2003). 'Crash, Theatre Audiences, and the Idea of "Liveness"'. *Studies in Theatre and Performance*, 23: 1, pp. 21–39.

Baudrillard, Jean. (1990). *Seduction*. Trans. Brian Singer. New York: St. Martin's Press.

Benjamin, Walter. (1969). 'The Work of Art in the Age of Mechanical Reproduction'. Trans. Harry Zohn, in *Illuminations*, ed. Hannah Arendt. New York: Schocken.

Blau, Herbert. (2003). 'Odd Anonymous Needs: The Audience in a Dramatized Society'. In *Critical Concepts in Literary and Cultural Studies: Performance*, Vol. II, ed. Philip Auslander. London, New York: Routledge, 269–81.

Bolter, Jay David and Richard, Grusin. (1996). 'Remediation'. *Configurations* 4: 3, pp. 311–358.

Breznican, Anthony. (2004). 'Spielberg, Zemeckis Say Video Games, Films Could Become One'. Associated Press, 4 September, http://SignOnSanDiego.com (accessed 22 March 2006).

McLuhan, Marshall. (1964). *Understanding Media: The Extensions of Man.* 2nd edn, New York: New American Library.

Pavis, Patrice. (1992). *Theatre at the Crossroads of Culture.* Trans. Loren Kruger. London, New York: Routledge.

Index

Abstände, 70
Ackroyd, Heather, 128
ADaPT, 49, 51, 57
'Adopt an Actor' Scheme, 181,
  189–91, 193
Agamben, Giorgio, 96, 99
AI (artificial intelligence), xv, 141, 143
Angelfish (dance with), 89, 91–8
apparatuses, interpretative, 10
apparatuses, technological, 14, 156
Artaud, Antonin, xvii, 18–30
authentic difference (in Bergson), 11
avatar, xv, 87, 141, 143, 147, 149

Baudrillard, Jean, 23, 196–7
Bausch, Pina, 32, 80, 82
  Der Fensterputzer, 80
becoming, 1–3, 8–11, 14, 20, 85–98,
  113, 116–22, 129, 149
becoming (Grosz), 1
'expert-becoming', 9
Bell, Clive, 127, 140
Bergson, Henri, xviii, 9, 10, 11, 13,
  116, 117, 121
  'le devenir' (becoming), 10
duration, durational, 1, 4, 7, 13–14
Berland, Elodie, 141, 144
Bessell, David, 142
biodiversity, 86
Birmingham Repertory
  Theatre, 181–4
Blakslee, Sandra, 149
Blast Theory, 51, 57, 58, 180
Blue Bloodshot Flowers, 141–9
'bodig', 8
body, as action in performance,
  11, 132
body, as container, 85
body, as corpse, 9
body, as effect, 11

'body', as measure, 9
body, 'heavenly', 9
Body Mind Centring, 114
'body politic', 9
body-screen, 100–3
Body Spaces, xix, 130, 174,
  177–9
'the body' in Spinoza, 1–16
Borradori, G., 11–12
Bourdieu, P., 4, 6
Bowden, Richard, 141–3, 150
'boys with toys', syndrome in
  interactive performing, 69
Breton, Andre, 18–19, 24–6
Bristol Old Vic Theatre, 181–3
Brown, Trisha, 76

Canguilhem, George, 154, 167
Carrel, Alexis, 167
causation, 132
Centres for Excellence in Teaching
  and Learning (CETL), 188
Chameleons Group, xvii, 18–29
  Chameleons 4: The Doors of Serenity,
  21–2, 27
Changing Room, The, 88, 92, 94, 98
choreography, writing spaces, 86
choreomediated, xviii, 100, 108
cinedance, 101–2, 106, 109–10
Claxton, Guy, 113, 114
cognitive decisions, 121, 123
'cognitive mapping' (in Jameson),
  4, 10
cognitive re-mapping, 10
comedy (and truth), 28–9
common sense, 2, 4
  common sensical, 7
community performance, 170
Company in Space, 46
complex systems, 116

consciousness, 25, 51, 113–23, 148
core consciousness, 114
pre-reflective consciousness, 113, 122
primary consciousness, 114
reflective consciousness, 113, 123
corporeal perception, xvi, 41, 43, 96, 107, 133, 148, 149, 175
Cronenberg, David, 52
eXistenZ, 52
'cross-modal transfer' (in Ulmer), 13
Cunningham, Merce, 46, 62
cybernetics, 45
cyborg, xix, 21, 169, 172, 174, 179

Damasio, Antonio, 113–14, 123–5
dance film, 31–42
Danks, Mark, 146
Daredevil, 179
Darley, Andrew, 46, 57
Dead East, Dead West, 141, 144–5, 147–9
DECface, 142, 143
Decouflé, Philippe, 46
DeLanda, M., 1–4, 11
'broken symmetry', 11
'immaterial becoming', 2
Deleuze, Gilles, xvi, 1–16, 91, 112, 116–17, 121
becoming, 3, 8, 9, 16
Kant's Critical Philosophy, xviii, 6, 98, 148
Spinoza: Practical Philosophy, 1–2
velocities, 8; differential, 8
Deleuze, Gilles and Guattari, Félix, xviii, 1, 2–6, 8–12, 98, 148
affecting and affected, 8
effect, effected, 10–12
logic (Deleuze), 2–6; of becoming, 3; of production, 6; of sense, 2
Anti-Oedipus, 1
'plane of immanence', 4, 11
A Thousand Plateaus, 1, 98
Deren, Maya, 32, 36
Derrida, Jacques, 19, 20
desperate optimists, 3
diagetic, 135, 137
Digital Cultures Lab, 78, 79
digital economy, 7, 8

digital environments, 91, 94
disability performance, 169–80
discipline, 4, 12, 14
disciplinary, 3, 4, 6, 12, 14, 16
Disembodied Cuisine, 162, 163
DIY De-victimizer, 165
Dupras, Martin, 144, 146
DV8, 3, 42
dynamogenesis, 131

Edelman, Gerald, 113, 114, 123
educational policy, 185, 188
Ekman, Rasmus, 135
electronic rhetoric (in Ulmer), 8
embodied interface, xviii, 87, 93, 97
embodied self, 149
emergence, xvii, 4, 10, 45, 133
emergent, 2, 4, 13, 44, 45, 57, 95, 96, 114, 129
environments, online, 43–59
derived, 46
immersive, 45
mixed-reality, 49
networked, 45
programmable, 44
sensory, 45
virtual, 43, 44, 92, 147
equipment (in Rabinow), 4, 10
Ettinger, Bracha L., 96
Evans, Maria, 186
EverQuest, 46
expert and expertise, 1–16
intuition, 12, 13
invention, 3
Exploring Shakespeare, 181, 185, 188, 191
Extended Body, 166
Extra Ear – 1/4 Scale, 159–61

Feeding Ritual, 156, 159
'feeling for the avatar', 94
Fuller, Loïe, xviii, 101–7

Gabor, Dennis, 135–6, 140
Gallese, Victor, 114
games (computer, online, video), xvii, 43–59, 114
GEM (Graphics Environment for Multimedia), 146

generative interaction, 146
Geoface, 142–3
Globe Theatre, 181–3, 189–91
government funding, 185
Granulab, 135–42
Grimson motion tracker, 143, 150
Guattari, Félix (and Deleuze), xviii,
   1–4, 98, 121, 148

Haffner, Nik, 46
Hall, Stella, 128, 140
Halo, 46
Hamlet, 185–7, 193
haptic, 52, 58, 107, 113
Hattosh-Nemeth, Jez, 144
Hayles, Katherine, N., xvi, 7, 10, 11,
   16, 56
HCI (human–computer
   interaction), 45
Hébert, Bernar, 31, 39, 40
Heidegger, Martin, 13
Henry V, 183
Higher Education Funding Council
   (HEFCE), 188
homo ludens, 48
Hubbard, Edward M., 148, 150
Huxley, Julian, 163, 168
hybrid, hybridity (in performance),
   49, 57, 75, 92, 147, 148
hypotyposis, schematic and symbolic,
   xvi, 15

impulse (performance), 131–4
infrared sensors, 47, 61
instantons, 134, 135
insults (and comedy), 27–9
'Intelligence, Interaction, Reaction
   and Performance', 141–50
interaction, artistic justifications, 73
interactive effect, 64, 71, 72
interiority, 14, 15
   'deep', 15
   'inside', 14, 15
intuition
   as 'attuned empiricism' (in Grosz),
      13–14
   as 'shadow' (in Bergson), 13
intuition, eidetic intuition, 131
Isobe, Katsura, 144–5

Jeremiah, xv, 141–50
Jones, Bill T., 46, 76
Jones, Robert Edmond, 195
Jousse, Marcel, 131
judgement, 4, 6, 16
   as measure, 4, 16
   of taste and value (in Bourdieu), 4, 6
Judson Church Dance Theatre, 115

Kalypso, 66, 76
killing ritual, 158–60, 163
kinaesthesia, 41, 106, 178
kinaesthetic, xviii, 63, 71, 91–3, 97,
   100–10, 113, 120, 169
Kwon, Miwon, 94

La La La Human Steps, 33,
   39, 42
Lecavalier, Louise, 39–41
Lefebvre, Henri, 94
LifeForms (software), 46
liminal, xix, 124, 141, 149, 150
locative media, 34
Lock, Édouard, 31, 40
Longstaff, Jeffrey, 144

Macbeth, 185–6
McClave, Brian, 144
McGregor, Wayne, 3, 46, 56
machinima, 44, 57
Magnin, Pascal, 31, 35
Maguire, Sara Domville, 175
Marey, Etienne Jules, xviii, 101
Marks, Laura, 171
Martin, John, 114
Massumi, Brian, xvi, 3, 10, 11, 59
   event, 10, 11
   qualitative transformation, 3, 9
materials-driven vs content-driven
   work, xviii, 81, 83
matrices, of becoming, 95
Matrix, The, 53
matrixial spaces, 85
measure, 1–16
   as intuitive (in Borradori), 12
   and judgement, 4, 12
   qualitative and quantitative, 3
memory (in Borradori), 12

Merleau-Ponty, Maurice, xv, 148, 170–9
metakinesis, 114
metaphor, xviii, 4–7, 20–1, 30, 51, 53, 85, 91, 92, 102, 148, 172, 185
'congealed metaphor' (in Hayles), 7
jarring, 148
'metric properties', 4
mixed reality, 46, 49, 57, 86
Mnouchkine, A., 3
mods, 44
'momentary instantiation' (in Knorr), 3
motion capture, defined, xv, 46, 61, 62, 149
motion sensing, defined, 53, 55, 62
motion tracking, xv, 45, 47, 61, 76, 78, 110, 144
motion tracking, defined, xv, 45, 47, 61
movement capture, 45
movement mapping, 102
Mullins, Aimee, xix, 170–5
multi-dimensional mapping, 67, 69
multiplicities, 2, 91, 98, 117

national curriculum, 182, 184, 188, 193
National Theatre, 181–4, 192
network (of values), 9
neuroscience, 112
neuroscientist, 113, 114
'new work', 2–6, 10, 14
Newman, James, 48
Nicoll, Allardyce, 23, 24, 195
Nietzsche, Friedrich, 20, 30
nominalisation, 8
and naming, 8
and uses of the noun and definite article, 8
nuance (in Bergson), 11

Obermaier, Klaus, 78
Apparitions, 78
objectification, 11
Olimpias, The, xii, 179
ontology, 20, 24, 196
ontologising, 7

Osborne, P., 3
outside of writing, 2, 11, 12

Palindrome Intermedia Performance Group, 60–76
and camera angles, 71–2
EyeCon (software), xvii, 50, 62, 64–5; 'dynamic field', 65; 'feature fields', 65
position tracking, 66, 69, 70, 71
remote sensors, 61
percept, 121, 122
performative triggers, 15
'phase transition' (in DeLanda), 1, 11
phenomenology, 110, 131, 134, 153, 156, 165, 169–80
physical/physiological sensors, 61
pitch, 66, 120, 124, 135–8
plasticity, 49, 50, 52
pleasure (of play), 48–9
Poincaré, Henri, 134, 140
posthuman, 10, 11, 46
'pre-posthuman', 3, 10
production, economy of, 4
logic of, 6
processes of, 3
proprioception, 50, 52, 122, 123, 133
proprioceptive, 43, 49, 91, 113
psychophysical, 116, 122
Puckette, Miller, 146

qualitative multiplicities, 117

Ramachandran, V.S., 148, 149, 150
real time, defined, 6, 41, 44, 56, 100, 105, 112, 146, 148
recognition, 44, 49, 50, 142
sudden recognition (in Ulmer), 13
relational, 8, 9, 14, 15, 85, 101, 108–9, 117
relational dynamics, 109
responsive systems, xviii, 113
rewards (in gaming), 56
Ritual Theatre, 129
Roads, Curtis, 135
Rodoway, Paul, 113
Royal Shakespeare Company, 181, 182, 185

rules, 44, 48, 51, 56
rule-based games, 48
Rylance, Mark, 190

Saira Virous, 43–59
Schiphorst, Thecla, 46, 125
screenography, 100, 101, 105
semiotics (in Peirce), 6
Sensuous Geographies, 112–26
Sharir, Yacov, 46
shifter, 8
'the body' as shifter, 8
Short Sighted, 128
signature, 3, 9, 14
practice, 3, 14
Sim City, 46
Smith, Dave, 144
Sobchack, Vivian, 171
sonic quanta, 136
Spawn, 87, 89, 92, 93, 98
Spottiswoode, Patrick, 190
Stagework, 181, 183, 186, 188, 191
Stanier, Phil, 141, 150
stereoscopic video processing, 147
surrealism, 19, 24, 26
Surrealist Manifesto, 18, 24, 26
synaesthesia, 52, 150

telematic, 43–59
tele-action, 56
telepresence, xvii, 45–6, 55–6, 135
theatre of ideas, 80
Theatre de Complicite, 3
'thing itself', 3
Thirteenth Floor, The, 53
Thornton, K.D., 157, 167
Tiernan, Terence, 145
time, historical, 9
'unhinge[d]', 7
Tissue Culture and Art Project, The
partial life, 154, 159, 160
Pig Wings, The, 158–60
Relifing Roadkill, 165
Semi-Living, 153–68
'Techno-Scientific Body', 154–5,
165–6
tissue engineering, xv, xix, 162, 163
Victimless Leather, 163–4

3D image manipulation, 146–7
trajets, 100–11
translation (historical impact), 1, 108
Troika Ranch
Isadora (software), 50, 58, 66
16 [R]evolutions, 78–82
Truax, Barry, 136, 140
'truth' (seeker), xvi, 18–30

Ulmer, G., 8, 13
'recentring insight', 13

Verity Smith, Paul, 144
Vez, Ultima, 31, 33
video sculptures, 44
video surveillance cameras, 60, 150
videodance, 10, 100, 106, 109
Virilio, Paul, 86, 90, 167
Viroid Flophouse, 51, 55
virtual other, 87–99
virtual philosophy (in DeLanda), 2–4
virtual reality, 29, 45, 52, 94
virtual world design, 44
visceral, 86, 107, 113, 122, 161, 165,
170–2, 179

Warr, Tracey, 129, 140
Waters, Keith, 150
webcast, 50
White, P.R., 167
Wilton, Tom, 144, 145
Winkler, Todd, 150
Wolek, Nathan, 138, 140
Wooster Group, the, 3
writing, 1–16
electronic, 44
expert, 11
orders of, 2
outside of, 2, 11, 12
performance, 2, 8, 9, 11, 14
of spaces, 86

Xenakis, Iannis, 135, 140
Xstasis, 138, 139

Zeman, Adam, 142
Ziegler, Christian, 46
ZKM Media Museum, 47